WINE QUALITY

Food Industry Briefing Series

WINE QUALITY: TASTING AND SELECTION

Keith Grainger

A John Wiley & Sons, Ltd., Publication

Blackwell Publishing was acquired by John Wiley & Sons in February 2007. Blackwell's publishing programme has been merged with Wiley's global Scientific, Technical, and Medical business to form Wiley-Blackwell.

Registered office
John Wiley & Sons Ltd, The Atrium, Southern Gate, Chichester, West Sussex, PO19 8SQ, United Kingdom

Editorial offices
9600 Garsington Road, Oxford, OX4 2DQ, United Kingdom
2121 State Avenue, Ames, Iowa 50014-8300, USA

For details of our global editorial offices, for customer services and for information about how to apply for permission to reuse the copyright material in this book please see our website at www.wiley.com/wiley-blackwell.

Library of Congress Cataloging-in-Publication Data

Grainger, Keith.
  Wine quality : tasting and selection / Keith Grainger.
    p. cm. — (Food industry briefing series)
  Includes bibliographical references and index.
  ISBN 978-1-4051-1366-3 (pbk. : alk. paper) 1. Wine tasting. 2. Wine and wine making–Analysis. I. Title.
  TP548.5.A5G73 2009
  641.2′2—dc22
                                                              2008030012

A catalogue record for this book is available from the British Library.

Set in 10/13pt Franklin Gothic Book by Aptara® Inc., New Delhi, India
Printed and bound in Malaysia by KHL Printing Co Sdn Bhd

1 2009

Contents

WINE QUALITY

CONTENTS

WINE QUALITY

A colour plate section appears between pages 28 and 29

CONTENTS

Series Editor's Foreword

Wine is surely a gift of the gods. If mankind had not been meant to enjoy wine, then why would the gods have created *Vitis vinifera* and its relatives, and why would we have been given the capacity to extract grape juice and exploit yeast in controlled fermentations to produce this most wonderful of beverages? Wine is an elegant drink, and more. What may appear to be a simple glass of wine can contain a complexity of aromas, flavours, textures and colours encapsulating the skills of generations of winemakers developed in many countries and on many continents. Wine is an intrinsic part of the history of mankind, and the pursuits of viticulture and vinification have helped to shape both cultures and societies. To deny the value of wine to the cycle of our lives is to reveal prejudice and ignorance, for good wine, consumed sensibly, has the power to augment the meal, bring joy to social occasions and enhance the quality of our lives. Yet though we may drink wine, how many of us actually taste wine and how many do so knowledgeably?

In Grainger's *Wine Quality* the author, Keith Grainger, explains how to taste wine and how to develop knowledge about wine. He explains how to go about understanding wine, how to make an informed and objective assessment of wine and, consequently, how to gain greater pleasure and interest from wine. Grainger's *Wine Quality* is logically organised. The book explains the different elements of objective of wine tasting and how to undertake the practices of assessment in an organised and professional way. It also describes product faults and their causes as well as giving comments on factors affecting the quality of wines and issues relating to production constraints, the marketability of wines and the markets in which wines are found. Here Grainger has distilled – or should I say, fermented? – many years of practical experience as a

WINE QUALITY

wine educator to those who would know more about wine, as well as a consultant to the wine industry. The book has been written as a companion to *Wine Production* by Keith Grainger and Hazel Tattersall (Blackwell Science). Together these publications provide the reader with a quick and easily accessible understanding of how wine is made and how it should be judged.

Grainger's *Wine Quality* and Grainger and Tattersall's *Wine Production* are both from the Blackwell Science *Food Industry Briefing* series of books. The series was created to provide the food industry with a resource for use by managers and executives who need to broaden their knowledge without devoting large amounts of time to study. It was also conceived as a resource for the development of staff to increase their expertise. Additionally, the *Food Industry Briefing Series* should be of benefit to lecturers in the fields of food science and technology and their students. Today, food industry professionals, lecturers, students and university libraries are all subject to tight budgetary controls. With this in mind, the *Food Industry Briefing Series* has been conceived as a source of high-quality texts that fall well below the price threshold of most technical and academic texts.

Ralph Early
Series Editor, *Food Industry Briefing Series*
Harper Adams University College
June 2008

Preface

When, in 1920, Professor George Saintsbury's *Notes on a Cellar Book* was first published, the 75-year-old author could have had no idea that sharing his opinions of wines he had drunk over more than half a century would be the birth of a new art, and the precursor of a new science, of the assessment of the tastes and quality of wines. Saintsbury was to become an icon to the oenophile, having both a prestigious wine and dining club, and a flagship Californian winery named in his honour. The clarets of 1888 and 1889 were, to Saintsbury, reminiscent of Browning's *A Pretty Woman* and the red wines of the south of France 'Hugonic in character'. Today's wine critic is, rightly or wrongly, usually more concerned with matters of acidity, balance, length and perhaps the awarding of points than with allusions to Browning or Hugo. Today, the professional taster also strives to be objective in the assessments made, something that Saintsbury would never claim to be.

This book aims to provide a concise, structured yet readable understanding of the concepts and techniques of tasting, assessing and evaluating wines for their styles and qualities, and of the challenges in assessing and recognising quality in wines. Also discussed are the faults that can destroy wines at any quality level and the misconceptions as to what constitutes quality. The book does not examine grape varieties in detail or, other than by way of example, the profiles and qualities of the vast array of wines produced in the wine regions of the world. There is already a wealth of literature on these topics. Nor do I cover the basics of viticulture and vinification, other than areas that particularly impact on quality. For this topic the reader is referred to *Wine Production: Vine to Bottle*, by Keith Grainger and Hazel Tattersall (published by Blackwell Science in the *Food Industry Briefing Series*).

WINE QUALITY

This book may prove valuable to wine trade students and professionals, sommeliers, restaurateurs, food and beverage staff and managers and all true wine lovers. The tasting structure and example tasting terms used herein are generally those of the Systematic Approach to Tasting of the diploma level of the *Wine & Spirit Education Trust*. As such, I hope the book may prove valuable to those studying for, or considering studying for, this qualification. Although the book is written primarily for the reader with limited scientific knowledge, at times it is necessary to take a more scientific approach, especially when examining the compounds that give rise to aromas, flavours and, particularly, taints. The text is also unashamedly interspersed with the occasional anecdote, for it is not just our personal perceptions but also our experiences that shape our interaction with what can be the most exciting of beverages.

I wish to thank everybody who has given me their time, knowledge and opinions. I also wish especially to thank Ralph Early, series editor for the *Food Industry Briefing* books, for his input and work in editing the text, Nigel Balmforth and Kate Nuttall of Blackwell Publishers for their support (and patience), Geoff Taylor of Corkwise for analysis information, Antony Moss for reviewing the, text and making valuable suggestions, and finally Hazel Tattersall for her work in preparing the index. I also wish to particularly thank the *Wine & Spirit Education Trust* for allowing me to use, adapt and extract from the WSET® Systematic Approach to Tasting (Diploma).

Keith Grainger

Acknowledgements

Figure 1.1	British Standards Institute
Colour Plate 2.3	Cephas/Diana Mewes
Colour Plate 2.4	Cephas/Diana Mewes
Colour Plate 2.7	Cephas/Alain Proust
Colour Plate 10.1	Rustenberg/Keith Phillips
Colour Plate 10.2	Rustenberg/Keith Phillips
All other colour plates and figures:	Keith Grainger

Acknowledgement is given to *Wine & Spirit Education Trust* for allowing the use of the WSET® Systematic Approach to Tasting (Diploma)

Introduction

How can wine quality be defined and how can it be assessed? These apparently simple questions pose many more. Wine styles are not static – in wine regions throughout the world, the wines made today are different to those of previous generations. Few would deny that the quality of inexpensive wine is higher than ever, but the paradoxical question is, are they *better* wines? And what of the so-called great wines from illustrious producers? Are they richer, softer and more voluptuous, or are they less distinctive, less a statement of place, less charming, less exciting even less enjoyable than the wines of a generation or two ago?

Tastes change and wine is, as it always has been, subject to the vagaries of fashion. In 1982, Master of Wine and Burgundy expert Anthony Hanson wrote in the first edition of his critically acclaimed book *Burgundy* 'great Burgundy smells of shit'. If there were any raised eyebrows at the time, these were only because of Hanson's choice of language – indeed many Burgundies had a nose of stables, farmyards and the contents thereof. By 1995 Hanson was finding such a nose objectionable, and laid the finger of blame at microbial activity. We now know that these smells have nothing to do with Pinot Noir (the variety from which pretty much all red Burgundy is made) or the Burgundy terroir, but stem largely from a rogue yeast: *Brettanomyces*. Today, *Brettanomyces* is generally regarded as a fault in wine (see Chapter 6).

So something that in 1982 was regarded by an expert taster as a sign of quality is today seen as a fault. We may draw a further example of mature Riesling. Producers in Germany and Alsace have long lauded the diesel or kerosene nose that these wines can exhibit after several years in the bottle. Many New World producers and wine critics regard such a nose as a flaw, caused by TDN

(1,1,6-trimethyl-1,2-dihydronaphthalene, also known as noriso-prenoid). In common with many other wine writers I disagree, finding such a nose part of the individual, sensuous character of this most distinctive of varieties.

There are apparent contradictions in how we assess and define wine quality. One wine can be analysed chemically and microbio-logically and be declared technically very good, yet may taste distinctly uninteresting; another wine may show technical weaknesses or even flaws, yet when tasted it may be so full of character and true to its origin that it sends a shiver down the spine. And how can we define what constitutes a truly great wine? In his book, *The World's Greatest Wine Estates*, Robert Parker, without doubt the world's most influential wine critic, gives a workable definition of greatness as:

(1)   'the ability to please both the palate and the intellect';
(2)   'the ability to hold the taster's interest';
(3)   'the ability to display a singular personality';
(4)   'the ability to reflect the place of origin'.

It is interesting that many wine lovers bemoan the 'Parkerization' of some wines – i.e. the sense of place is negated as producers strive to produce the style of wine driven by super-ripe fruit that they believe will earn them a high Parker rating.

The very concept of assessing quality in food or drink is something that does not come easily to many people. To them a quality product is one with a 'designer' label, a well-known brand, advertised on television or in glossy magazines and which is in fashion. In other words, something that they are told is good by believable sources. The British education system, as it is now institutionalised in the United Kingdom, is bound by the straightjacket of the National Curriculum and held hostage by the need to meet targets in assessments. Consequently, it completely fails to encourage young people to develop the life-enhancing skills of discerning quality. Pupils may be taught about food and nutrition but leave school unable to distinguish the fine from the mediocre: to the benefit of the many food businesses that make a lot of money from second-rate food products. It is of great concern to many wine producers that so many 'twenty-something' consumers are not turning on to wine in the way that the previous two generations did, and most of those that do seem unwilling to leave the simplistic world of 'entry-level' quaffing liquid.

Quality may be regarded as an objective standard of excellence, with an absence of any faults. This leads us to another major issue to challenge tasters: objectivity. Objectivity is generally regarded as seeing something as it really is, uncoloured by personal preference or bias. A tasting assessment should be structured so that the taster perceives a wine to be as it truly is. However, is this achievable? The argument as to whether or not there is such thing as objectivity has induced growing perplexity during recent decades. Is objectivity the seeing of reality, something actually existing? Or is it simply taking pains to diminish or eliminate bias? Should we distinguish between ontological objectivity (seeing things as they truly are, the truth coinciding with reality) and procedural objectivity (using methods designed to eliminate personal judgements, perhaps coloured by feelings or opinions)? The key question is whether we can know if our views of reality actually correspond with it, and just how do we represent our views of reality?

In order to try and ensure rigour, validity acceptance and realism of findings, researchers using the methods of the natural sciences use procedures that endeavour to eliminate subjectivity, which may be described as procedurally objective methods to gain an ontologically objective understanding. When tasting wines it is important that we use techniques that are procedurally objective but realise that our assessment is not an ontologically objective one. In other words it is a judgement and as such it is fallible.

Judgements as to quality are, of course, framework dependent. Frameworks include those appertaining to the taster and to the wine. A taster, however open-minded he or she tries and claims to be, will work within boundaries established by training, history and culture. A Burgundian winemaker trained at the University of Dijon who has worked, lived and breathed the terroir-driven wines of the Côte d'Or will, however well travelled and widely experienced, assesses a powerful fruit-driven Chardonnay from Napa Valley very differently to a UC Davis trained American oenologue. A Muscadet de Sèvre-et-Maine AC – Sur Lie, however well crafted and showing *tipicity de-luxe* with classic autolytic character would, by the vast majority of trained tasters, not even be placed in the same quality league as a Montrachet Grand Cru AC. Yet both might be wonderfully enjoy-able, just as simple cod and chips (which would accompany either wine) might please the diner as much as lemon sole prepared by the chef of a Michelin three-star restaurant. And, of course, the more illustrious the origin and producer of a wine, the higher the price, the greater the expectations of quality, and the deeper the disap-pointment should it under perform. In other words for both the pro-ducer and the consumer, quality is not something that can simply be bought or that can always be relied upon. Even a subtle change in any of the multifarious variables that constitute the make-up of a wine will impact, positively or negatively, on taste and quality. But this is just one of the factors that make the tasting and assessment of wines so exciting.

Wine Tasting

The history of winemaking goes back some 8000 years, which means that the history of wine tasting, at least in a basic way, is just as old. References to the taste of wine abound in works through the centuries. On 10 April 1663, the diarist Samuel Pepys wrote that he drank at the Royal Oak Tavern 'a sort of French wine called *Ho Bryen*, that had a good and most particular taste I ever met with'. Pepys's note might not have been sufficient for a pass in today's wine trade examinations, but he had the disadvantage, or should that be benefit, of not have been inundated with press releases or the pronouncements of wine writers, critics and sommeliers. He tasted the wine, and gave his perceptions of it.

1.1 Wine tasting and laboratory analysis

There are two basic ways by which wines may be analysed: by scientific means using laboratory equipment and by the organoleptic method, i.e. tasting. A laboratory analysis can tell us a great deal about a wine, including its alcohol by volume, the levels of free and total sulfur dioxide, total acidity, residual sugar, the amount of dissolved oxygen, and whether the wine contains disastrous spoilage compounds such as 2,4,6-trichloroanisole or 2,4,6-tribromoanisole. It is highly desirable that producers carry out a comprehensive laboratory analysis both pre- and post-bottling. If another laboratory undertakes a duplicate analysis, the results should be replicated, allowing for any accepted margins of error. Scientific analysis can also give indications as to the wine's style, balance, flavours and quality. However, it is only by tasting a wine that we can determine these completely and accurately. If a team of trained tasters assess the same wine, they will generally each reach broadly similar conclusions, although there may be dissention on some aspects, and occasionally out and out dispute.

Wine is, of course, a beverage made to be drunk and (hopefully) enjoyed. Low-priced wines are usually, at best, little more than pleasant, fruity, alcoholic drinks. As we move up the price and quality scale, wines can show remarkable diversity, individuality and imitable characteristics of their origin. Good quality wines excite and stimulate with their palettes of flavours and tones, their structure and complexity. Fine wine can send a shiver down the spine, fascinate, excite, move and maybe even penetrate the very soul of the taster. No amount of laboratory testing can reveal these qualities. Further, it is only by tasting that the complex intra- and interrelationship between all the components of the technical make-up of a range of wines and human interaction with these can be truly established. It can be argued that the perceptions of the taster are all that really matter – wine is not made to be tasted by machines, but by people.

1.2 What makes a good wine taster?

Developing wine-tasting skills is not as difficult as many would imagine. Whilst it is true that some people are born with natural talent (as with any art or craft), without practice and development such talent is wasted. People who believe that they will not make good wine tasters due to a lack of inborn ability should perhaps ask themselves some simple questions: Can I see, smell and taste the difference between oranges, lemons and grapefruit, or between blackcurrants, blackberries and raspberries? If the answer is yes, the door is open. There are a few people, known as anosmics, who have a poor or damaged sense of smell, and obviously they are unable to become proficient tasters, and a larger number of people who are specific anosmics, i.e. lacking the ability to detect certain individual aromas. It is also true that some people have on the tongue a high density of fungiform papillae, which contain the taste buds, making them particularly sensitive to bitter sensations.

It has been argued by Yale University Professor Linda Bartoshuk that this group of people are 'supertasters'. Ann Noble's group at UC Davis has also established that there are no 'supertasters in general', but that an individual who is a supertaster with one bitter compound, e.g. naringine, might be a non-taster with another, e.g. 6-N-propylthiouracil or caffeine. It should be noted that supertasters do not necessarily make the best wine tasters, for the intense sensations they perceive from bitterness and astringency impacts on other sensations and perceptions of the balance of the wine.

With practice and concentration, the senses needed for wine tasting can be developed and refined. Memory and organisational skills also need to be developed: it is not of much use having the sensory skills to distinguish between, say, an inexpensive young, Cabernet Sauvignon from Maule (Chile) and a fine mature Merlot-dominated wine from Pomerol (Bordeaux, France) if one cannot organise the characteristics in the brain and remember them. Thus the making of detailed and structured tasting notes is important – the very act of noting observations sharpens perceptions, and maintaining a consistent structure enables wines to be assessed, compared and contrasted. However, applying verbal descriptions to complex and possibly individual aroma and flavour perceptions poses many challenges. Learning too is important, for the taster needs to understand the reasons for the complex aromas and flavours and be able to accurately describe them. In short, there is no substitute for the widest possible tasting experience, encompassing wines of all types, styles, qualities, regions and countries of origin.

When tasting wines we are using the senses of sight, smell, taste and touch. The sense that requires the most development is that of smell. Smells create memory. You can walk into a room and, in an instant, you are reminded of another time and place – perhaps back in your infants' school classroom or in grandma's house. In the briefest of moments your nose has detected the constituents, analysed them and passed the information to the brain which has immediately related them to a point in the memory bank.

For most people it is not difficult to develop the sense of smell. We live in a world in which we are conditioned to believe that many everyday smells are unpleasant and thus we try to ignore them. Walking in a city centre we may be subjected to a melange of traffic fumes, yesterday's takeaways and detritus of humankind and are tempted, even programmed by the media and society, to try and ignore the onslaught. Smells may be attractive or repulsive, and an attractive smell to one person may not be to another. The smells of the human body are a key component of attraction, sexual or general, or of rejection. Animal smells in particular are offensive to many – to say that somebody smells like a dog, horse or mouse would hardly be considered a compliment!

A simple way to help develop the sense of smell is to use it. When walking into a room smell it, smell the newly washed laundry, the material of clothes on a shop rail, the hedgerow blossom, even the person standing next to you. And, most importantly, commit these to memory. Expert wine tasters structure and organise a memory bank of smell and taste profiles and thus can relate current experiences to similar ones they have encountered. Interestingly, research by Castrioto-Scanderberg *et al.* (2005) using brain monitoring by means of functional magnetic resonance imaging shows that experienced tasters have additional areas of the brain activated during the tasting process, namely the front of the amygdala-hippocampal area, activated during the actual tasting and the left side of the same area during the aftertaste (finish) phase.

1.3   Where and when to taste – suitable conditions

The places that wines may be tasted are perhaps as diverse as wines themselves, and even less than technically ideal situations can have advantages. There is something magical about a tasting conducted in the vineyard, and moving from barrel to barrel in a producer's cellar can fill one with a real sense of time and place. On the other hand, exhibitions and trade shows, in spite of all the discomfort, noise and other distractions, can present a good opportunity to compare and contrast a large number of wines in a very short space of time.

However, for a detailed organoleptical analysis of wines an appropriate tasting environment is required, and the ideal tasting room will have the following characteristics:

Large: Plenty of room is necessary to give the taster his or her personal space and help concentrate on the tasting.

Light: Good daylight is ideal, and the room (if situated in the northern hemisphere) should have large, north-facing windows. If artificial light is required the tubes/bulbs should be colour corrected in order that the true appearance of the wines may be ascertained.

White tables/surfaces: Holding the glasses over a white background is necessary to assess the appearance and show the true colour of the wine, uncorrupted by surrounding surfaces.

Free from distractions: Extraneous noises are undesirable and smells can severely impact on the perceived nose of the wines. Tasting rooms should not be sited near kitchens or restaurants – an amazing number of New World wineries fail to have regard for this. Tasters should avoid wearing aftershaves or perfumes, and obviously smoking should not take place in the vicinity. There is no doubt that building materials, decorations furnishings and people all exude smells. Indeed, identical wines can be perceived differently according to the surroundings in which they are assessed.

Sinks and spittoons: Spittoons are essential (see below) and sinks for emptying and rinsing glasses are desirable.

As to when to taste, the decision is unfortunately often dictated by matters beyond the taster's control. However, the ideal time is when the taster is most alert and the appetite stimulated – namely in the late morning. After a meal is certainly not the best time, for not only is the taster replete and perhaps drowsier (as all early-afternoon seminar presenters know), but the palate too is jaded and confused after the tastes of the food.

1.4 Appropriate equipment

Having appropriate equipment for the tasting is most important. This includes an adequate supply of tasting glasses, water, spittoons, tasting sheets for recording notes and, at a formal sit-down event, tasting mats.

1.4.1 Tasting glasses

It is important to taste wines using appropriate glasses. Experts do not universally agree as to the detailed design of the ideal tasting glass, but certain criteria are essential. These are listed in Table 1.1.

Two of the key characteristics are as follows:

Fine rim: A fine rim glass will roll the wine over the tip of the tongue, whilst an inexpensive glass with a beaded rim will throw the wine more to the centre. The tip of the tongue is the part of the mouth where we most detect sweetness.

Cup tapering inwards: The cup of the glass must taper inwards towards to top. This will develop, concentrate and retain the nose of the wine, and also facilitate tilting the glass and swirling the wine. It should be noted that cut glass is not appropriate for wine tasting, as it is impossible to ascertain the true depth of colour.

Glasses manufactured to the ISO tasting glass specification (ISO 3591) are very popular amongst many serious wine tasters, both professional and amateur. The specification of the ISO glass is shown in Fig. 1.1 and a photograph in shown in Fig. 1.2.

The ISO tasting glass is particularly good at revealing those faults perceptible on the nose, as detailed in Chapter 6. Whether they are the best glass for tasting particular wine types is very much

Table 1.1 Criteria for a suitable tasting glass

Clear glass
Minimum 10% 'crystal' content
Stem
Fine rim
Cup tapering inwards towards top
Minimum total capacity = 21 cL (approximately 7 fL oz)

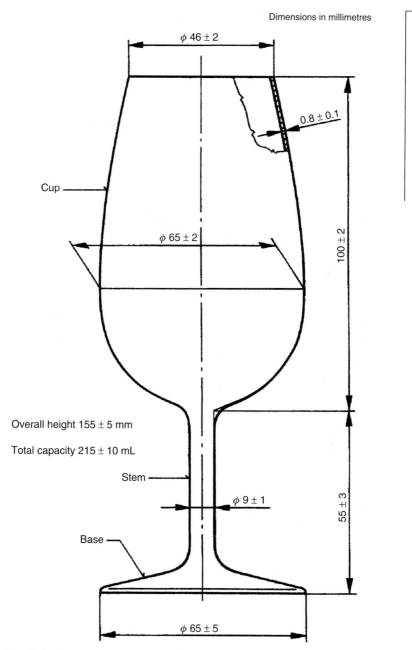

Dimensions in millimetres

φ 46 ± 2

0.8 ± 0.1

Cup

φ 65 ± 2

100 ± 2

Overall height 155 ± 5 mm

Total capacity 215 ± 10 mL

Stem

φ 9 ± 1

55 ± 3

Base

φ 65 ± 5

Fig. 1.1 ISO tasting glass specification

WINE QUALITY

Fig. 1.2 ISO tasting glass

open to discussion. The nose of full-bodied and complex red wines develops more in a larger glass; the Pinot Noir variety is more expressive in a rounder-shaped cup. Wine glass manufacturers, particularly *Riedel*, have designs to bring out the best of individual wine types, so perhaps the real advantage of the ISO glass is that it is a standard reference. However, the reader should be aware that some glasses marketed as ISO specification are definitely not having such deviations as beaded rims, larger cups or inferior soda lime manufacture.

An appropriate tasting sample is 3–4 cL, which will be sufficient for three or four tastes. At a formal sit-down tasting of a number of wines, pouring 5 cL into the glasses gives an opportunity to return for a further taste of the wine to see if there has been development in the glass and to compare and contrast with the other wines tasted. If glasses larger than the standard ISO glass are used, it is appropriate to pour correspondingly more wine.

Flutes – the ideal glasses for sparkling wines

Tall flutes are ideal for assessing sparkling wines. They should be fine rimmed and preferably with a crystal content. A tasting sample comprises a quarter or third of the capacity of the flute. The quality of the mousse (sparkle) is most clearly seen, and even the most delicate nose of the wine is enhanced. Interestingly, the method of manufacture of the glass makes a considerable difference to the size, consistency and longevity of the mousse (bubbles) in a sparkling wine. Handmade glasses give the most consistent bubbles of all, but any flute can be prepared to give a livelier mousse by rubbing some fine glasspaper on the inside of the bottom of the cup, immediately above the stem.

Glass washing and storage

Ideally, wine glasses should be washed by hand just in hot water. If the glasses show signs of grease or lipstick a little detergent may be used. The glasses should be well rinsed with hot water, briefly drained then dried using a clean, dry, glass cloth that has been previously washed without the use of rinse aid in the washing cycle. Glass cloths should be changed regularly – perhaps after drying as few as six glasses. The odour of a damp or dirty glass cloth will be retained in the glass and impact on the content. At an exhibition or trade tasting where the participants collect a glass from a collection on a table, the empty glass should always be nosed to check for basic cleanliness and absence of 'off' aromas.

Glasses should not be stored bowl down on shelves, for they may pick up the smell of the shelf and develop mustiness. Obviously, standing glasses upright on shelves may lead them to collecting dust, so a rack in which glasses are held upside down by the base on pegs is perhaps ideal.

In order to be sure that no taint from the glass is transmitted to the wine, it is a good idea to rinse the glass with a little of the wine to be tasted. This is also useful if tasting a number of wines from the same glass.

1.4.2 Water

There should be a supply of pure, still mineral or spring water for the taster to refresh the palate between wines, if necessary, for drinking and perhaps rinsing glasses. The variable amount of chlorine contained in tap water usually makes this unsuitable. Plain biscuits such as water biscuits may also be provided, but some tasters believe that these corrupt the palate a little. Cheese, although sometimes provided at tasting events, should be avoided as the fat it contains will coat the tongue and the protein combines with and softens the perception of wine tannins.

1.4.3 Spittoons

Spittoons, placed within easy reach of the participants, are essential at any serious wine tasting. Depending on the number of attendees and the capacity required, there are many possibilities. The simplest improvised spittoons are simply wine-cooling buckets, perhaps lined with sawdust or shredded paper in order to reduce splashing. There are many designs of purpose-built spittoons suitable for placing on tables and larger units for standing on the floor. Consideration should be given to the construction material: plastic, stainless steel and aluminium are all good. Unlined galvanised metal should be avoided at all cost as wine acids can react and create disgusting aromas. The importance of spitting at wine tastings cannot be over-emphasised, not least because the taster needs to keep a clear head and generally avoid unnecessary ingression of alcohol. Even when wines are spat out, a tiny amount will still make its way to the stomach, and indeed a minute amount will also enter the body via the act of nosing the wines.

1.4.4 Tasting sheets

Without doubt, making notes about the wines tasted is essential. Depending on the circumstances the notes may be brief or detailed for personal use only or for sharing or publication. In order to facilitate note taking, tasting sheets should be prepared, listing

---

Bodegas Trapiche S.A.
www.trapiche.com.ar

1. 2005 Trapiche Oak Cask Chardonnay
    Mendoza (C)
    9 Months in French & American oak barrels

6. 2004 Trapiche Malbec Single Vineyard
    Gei Berra Mendoza (E)
    18 months in new French oak barrels

2. 2004 Trapiche Oak Cask Pinot Noir
    Mendoza (C)
    9 Months in French & American oak barrels

7. 2004 Trapiche Malbec Single Vineyard
    Victoria Coletto   Mendoza (E)
    18 months in new French oak barrels

3. 2004 Trapiche Oak Cask Cabernet
    Sauvignon Mendoza (C)
    12 Months in French & American oak barrels

8. 2004 Trapiche Malbec Single Vineyard
    Pedro Gonzalez Mendoza (E)
    18 months in new French oak barrels

4. 2004 Trapiche Oak Cask Malbec
    Mendoza (C)
    12 Months in French & American oak barrels

9. 2004 Trapiche Medalla
    (Cabernet Sauvignon) Mendoza (E)
    18 months in new French oak barrels

5. 2004 Trapiche Broquel Cabernet
    Sauvignon Mendoza (D)
    15 months in new French & American oak
    barrels

10. 2004 Trapiche Iscay
     (Malbec & Merlot) Mendoza (E)
     18 months in new French oak barrels

Retail Price Categories: A = £4.99 & under   B = £5.00 - £5.99   C = £6.00 - £7.99   D = £8.00 - £9.99   E = £10.00+

Fig. 1.3 A simple tasting sheet

and detailing the wines to be tasted, with space for the participants to make notes. Background and technical analysis information can also be useful, either on the tasting sheet or as a separate handout. A simple tasting sheet as might be used at an exhibition tasting is shown in Fig. 1.3.

Use of tasting software

Software has been developed that facilitates the making of tasting notes on a hand-held computer, mobile phone or BlackBerry® device. The programme allows for consistently structured and detailed records to be maintained on an online database, unique to each user. Organisers of major trade tastings can submit details of the wines to be tasted to the software provider who makes the information available for download to the mobile device.

1.4.5   Tasting mats

If a number of wines are to be assessed at a formal sit-down tasting, each wine should have its own glass, placed on a paper tasting mat printed with circles of a size similar to the bases of the tasting glasses, each circle numbered and corresponding to the listed order of the wines on the tasting sheets. A simple tasting mat is shown in Fig. 1.4.

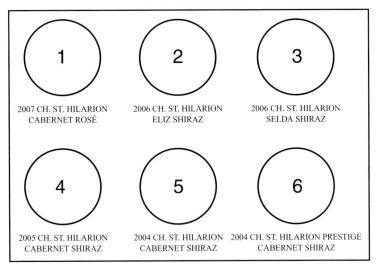

1	2	3
2007 CH. ST. HILARION CABERNET ROSÉ	2006 CH. ST. HILARION ELIZ SHIRAZ	2006 CH. ST. HILARION SELDA SHIRAZ

4	5	6
2005 CH. ST. HILARION CABERNET SHIRAZ	2004 CH. ST. HILARION CABERNET SHIRAZ	2004 CH. ST. HILARION PRESTIGE CABERNET SHIRAZ

Fig. 1.4   Tasting mat

1.5   Tasting order

If there are many wines to be tasted, of varying styles and qualities, it is sensible to do so in a considered order. There are several guidelines, but unfortunately many of these conflict:

- Sparkling wines should be tasted before still
- White wines should be tasted before red (but see below)
- Dry white wines should be tasted before sweet
- Light-bodied wines should be tasted before full-bodied
- Wines light in tannin should be tasted before those with high levels of tannin
- Young wines should be tasted before old
- Modest quality wines should be tasted before high quality

It will easily be seen that trying to sort a sensible tasting order for a wide range of wine styles and qualities proves challenging, especially as the characteristics of each wine may not be as anticipated. At a trade exhibition, tasting well in excess of 100 wines is not uncommon and even the most experienced tasters can suffer fatigue. It can be particularly difficult to taste a large number of sparkling wines, as the high acidity numbs the palate. Also, challenging to a taster is a large volume of red wines that are high in palate-numbing tannins. Many Master of Wine tasters advocate tasting red before white. The acid attack of whites makes red tannins seem more aggressive. The aromas of white wines are easy to assess after red wines.

1.6   Temperature of wines for tasting

The temperature of wines presented for tasting analysis is not necessarily that at which we would wish to drink them. This is particularly true of white wines, which many people prefer to drink relatively cold, perhaps 8–12°C (46–54°F). Coldness numbs the palate and white, rosé and sparkling wines are best tasted cool at 12–15°C rather than cold. Conversely, many people drink red wines at a relatively warm temperature. The expression 'room temperature' does not mean perhaps 22°C or more, the temperature of many a living room. The French language has an expression *chambré* which refers to bringing wines from cellar (storage) to serving temperature. Red wines are best tasted at 16–18°C, with the lighter reds and those made from Pinot Noir at the lower end of the scale. Some people might prefer to drink full-bodied reds a degree or two warmer than this, but tepid wines are distinctly unappetising.

1.7 Tasting for specific purposes

The way a tasting is approached and the type and detail of notes made may vary according to the purpose of the tasting and agenda of the taster. A supermarket or merchant buyer needs to consider marketability, consumer preferences, how a wine relates to others on the list and price point. An independent merchant selling fine wines may pay high regard to how true a wine is to its origin, often referred to in the wine world as *typicity*. A restaurateur particularly needs to have mind to match the wine with dishes on the menu. A winemaker choosing and preparing a blend looks beyond the taste of the component wine in the glass to the contribution it might make, in variable amounts, to a finished blend. This said, it is important for the taster to assess each wine as completely and objectively as possible by adopting a consistent and structured tasting technique as detailed below.

1.8 Structured tasting technique

Most people do not really taste wine, they simply drink it. But by undertaking a detailed and considered tasting ritual, all that the wine has to offer, good and bad, is assessed. Four headings should be considered in tasting each wine: **appearance, nose, palate** and **conclusions**. We examine our approach to these very briefly here, and in detail in Chapters 2–5.

1.8.1 Appearance

The appearance can tell much about a wine and give indications as to origin, style, quality and maturity as well as revealing some

Fig. 1.5 Wine glass tilted to assess appearance. To see a colour version of this figure, please see Plate 1 in the colour plate section that falls between pages 28 and 29

possible faults. The appearance of the wine should be examined in several ways, particularly by holding the glass at an angle of approximately 30° from the horizontal over a white background – perhaps a tablecloth or sheet of white paper. Such an assessment is shown in Fig. 1.5. This will enable the **clarity**, the **intensity** of colour and the true **colour** of the wine to be seen, uncorrupted by other colours in the room. Looking straight down on a glass of wine standing on a white background is also useful for determining the intensity of colour. Other observations should be made: e.g. any legs or tears that run down the surface of the glass, which are best seen by holding the glass at eye level and giving a gentle swirl to coat the wall of the cup.

1.8.2 Nose

A wine should be given a short sniff to check its **condition**. Most faults will show on the nose, and if the wine is clearly out of condition, we probably will not wish to continue past this point. Also, some of the more delicate volatile compounds are most easily ascertained by a short, gentle sniff or two. Further, the longer we are exposed to aromas, the less sensitive we become to them, so these first gentle sniffs are all important. Assuming the wine not to be faulty, it should now be aerated in order to help release the volatile aromas – i.e. putting air into the wine will make the nose more pronounced. The usual way of achieving this is by swirling the wine round the glass several times. This skill is quickly learned, but should you experience difficulty, the glass can be steadied on the side of a loosely clenched fist as you swirl. If the wine seems very dumb, the glass – appropriately covered – can be shaken vigorously for a second or two (this is seldom necessary other than for the very poorest of wines) and then the wine given several short sniffs. Try placing the nose at various points of the glass to see if the aromas are more pronounced or different. Very long sniffs should be avoided at this stage too, on account of the numbing effect. We should note the **intensity** of the nose – put very simply, how much smell do we get from the wine? The **development** of the nose, explained in detail in Chapter 3, will indicate the present stage of the wine in the maturity cycle. Most importantly we should analyse the **aroma characteristics**.

1.8.3 Palate

A novice watching an accomplished and experienced taster at work will perhaps be unsure as to whether to view the ritual with laughter, derision or wonder. Observing the taster nose the wine in detail may already raise eyebrows, but watching the subsequent ritual of slurping and chewing the liquid may seem like overacted theatre. However, a simple exercise will convince even the most sceptical of the value of a professional approach. A small sample of wine should be poured into a glass, drunk as one would normally drink and then reflected on for a moment. Then another mouthful, 1 cL is an appropriate amount, should be taken and assessed using the professional approach. It is important to take a suitable quantity of wine so it is not over-diluted by saliva and there is sufficient to assess it fully. The wine should be rolled over the tip of the tongue and air should be breathed into the wine. This is not difficult. The lips should be pursed with the head forward as air is drawn into the wine. The taster should not be concerned about the slurping noises made during this operation and forget the childhood scolding given by mother. Now the liquid should be thrown around the mouth – over the tongue, gums and teeth and into the cheeks. The wine should be chewed, making sure that the sides and back of the tongue are covered, and a little more air taken in. It will take 20 seconds or so to give a thorough assessment, but there is no point in retaining the wine in the mouth for longer than this as it will be diluted by saliva and the palate will have become numbed. Finally, the wine should be spat out and the taster should breathe out slowly and reflect. The huge number of sensations experienced during this exercise compared with simply drinking the wine will astound. The purpose of breathing air into the wine is to facilitate the vapourisation of the volatile compounds that travel via the retro-nasal passage to be sensed by the olfactory bulb.

The tip of the tongue will detect the level of **sweetness,** and the sides of the tongue and cheeks the **acidity**. Other areas of the tongue also detect these sensations, and this is discussed in Chapter 4. **Tannins,** normally only really relevant when tasting red wines, will be sensed particularly on the teeth and gums. The level of **alcohol** is felt as a warming sensation, especially on (but not limited to) the back of the mouth. The **body** of the wine is the weight of wine in the mouth. When tasting sparkling wines the entire mouth will also feel the sensations of the **mousse**. For the **flavour intensity** and **flavour characteristics,** it is not just the mouth at work but also the olfactory bulb which will receive volatile compounds via the retro-nasal passage. Finally, the all-important

length of the wine is the amount of time the flavours are retained on the palate after it has been spat out.

1.8.4 Conclusions

Having thoroughly assessed the wine, judgements and conclusions may now be made. The key consideration is the **quality**. Assuming one is tasting finished wines, the **price** or at least the **price category** should be determined – of course, the relationship between quality and price is the key to assessing value. The **readiness for drinking** should also be decided upon. If the wine is being tasted totally blind, that is without the taster knowing previously what the wine is, the vintage, the district, region or even country of origin, now is the time to mentally collate the information obtained during the structured tasting and reach a conclusion on these points.

1.9 The importance of keeping notes

Making, organising and keeping structured tasting notes is essential to improving tasting technique, to enabling wines tasted over time to be compared and contrasted and to providing a source of reference. The amount of detail included in the notes will obviously depend on the circumstances of the tasting, the time available, and the taster's specific focus and requirements. It is important to avoid any possible subsequent ambiguity or misunderstanding. This is vital if the notes are not intended for the taster's private use or are to be made for later publication.

The following chapters cover in more detail the structured tasting technique, the headings under which we consider the wines and make notes and also detail some appropriate descriptors. The tasting structure and example tasting terms used herein are generally those of the Systematic Approach to Tasting of the diploma level of the *Wine & Spirit Education Trust*. There are, of course, many other tasting expressions that the taster will wish to use, and the terms detailed in the ranges that follow are far from exhaustive.

However, unless the notes are purely for their own use, the taster is cautioned against using terms that are particularly personal to him or her. A note that the nose of the wine is 'reminiscent of Aunt Edna's lounge' will mean nothing to the reader who has not visited her. The retention and subsequent review of at least a selection of one's tasting notes, perhaps transcribed into notebooks or inputted into a computer, not only helps to develop and refine technique but also provides a reference library of wines' aromas and flavours.

Appearance

To many general wine drinkers and novice tasters, the detailed examination by a professional of a wine's appearance seems something of a pointless exercise. However, the appearance reveals much about a wine, giving indications as to style, quality, maturity and revealing some possible faults. In assessing the appearance of a wine we consider the following:

- Clarity
- Intensity
- Colour
- 'Other observations'

2.1 Clarity

A wine that has finished fermentation and has been stabilised for sale should be clear. Brightness is a sign of healthiness. Brightness is how light is reflected off the surface of a wine, which is related to the pH. Clarity is how light is scattered as it passes through the body of the wine which is related to turbidity, i.e. suspended particles. If a bottle containing sediment has been carelessly handled, then there may be fine or larger particles in suspension, but any finished wine that appears cloudy, oily, milky or otherwise murky is probably faulty. Possible faults are detailed in Chapter 6.

We may consider the clarity of a wine on the following scale:

bright — clear — dull — hazy

Wines from the New World are often brighter than those from Europe. Wines that have high acidity, particularly if tartaric acid has been added in the winemaking process, may appear to be especially bright. A young wine that appears dull probably has a high pH (low acidity), which is indicative of poor quality and total lack of ageing potential. As wines mature they lose brightness, and with over-maturity they become dull. A decrepit wine will look very flat.

2.2 Intensity

The intensity, i.e. the depth of colour, should be noted. An impression of the depth of colour can be obtained by looking directly down onto the wine in a glass that is standing on a white background. Fig. 2.1 shows a 6-year-old Amarone della Valpolicella DOC viewed from above. A more detailed examination is obtained by tilting the glass, again over a white background, and looking particularly at the heart of the wine. Fig. 2.2 shows the Amarone viewed in this way. It is important when comparing and contrasting several wines to ensure that the glasses contain the same quantity.

We may consider the intensity of a wine on the following scale:

water white — very pale — pale — medium — medium-deep — deep — opaque

Generally speaking, wines that are pale in colour will be lighter in flavour and body than those that are deeper coloured, but this is not always the case. Whites that have had extensive barrel ageing will have a greater intensity. Paleness in a red wine, especially a young wine, is generally indicative of a lack of concentration, perhaps as a result of high-yielding vines, or a cool climate. Some red grape varieties, particularly Pinot Noir and to a lesser extent Nebbiolo, are usually not very deep coloured, although there are some notable exceptions to this, e.g. some Pinot Noirs from New Zealand's Central Otago region. Full-bodied and concentrated reds, especially those from hot climates, in youth will be deep or opaque, and this suggests intense flavours and may be indicative of good quality. However, occasionally amazingly deep-coloured red wines can be surprisingly light in flavour. There

Fig. 2.1 Looking down for an impression of intensity. To see a colour version of this figure, please see Plate 2 in the colour plate section that falls between pages 28 and 29

are a few red varieties that have coloured flesh as well as black skins (e.g. Alicante Bouschet), but which are surprisingly light in flavour and these *teinturier* grapes may be used in a blend to deepen the colour. Further, very deep colours can sometimes indicate winemaking methods that have focused on colour, rather than flavour extraction. Mindful of the association in the eyes of drinkers (and critics) of deep red wines and perceived high quality, winemakers can manipulate intensity, for example, by the addition of *megapurple*, a concentrate made from grape skins and seeds. Thus, the taster is cautioned against drawing conclusions at this stage.

Fig. 2.2 The same wine as Fig. 2.1 viewed tilted to 30°. To see a colour version of this figure, please see Plate 3 in the colour plate section that falls between pages 28 and 29

2.3 Colour

Many factors affect the colour of a wine, including climate and region of production, grape variety or varieties, grape ripeness, vinification techniques including any barrel ageing and maturity.

2.3.1 White wines

White wines can vary in colour from almost water clear to deep gold or even amber. We may consider the colour of a white wine on the following scale:

water clear — lemon green — lemon — gold — amber — brown

Some white varieties such as Sauvignon Blanc generally produce wines at the lemon-green end of the range, while others such as Gewürztraminer (which has a more heavily pigmented skin) give straw or gold colours. Whites from cooler climates often appear lemon-green or lemon, and those from warmer areas straw or even gold. Sweeter wines such as Sauternes AC are usually gold, even in youth, but all whites will darken with age. The rate at which this takes place varies considerably depending on several factors, particularly acidity (higher acidity slows down the process) and how well the bottle has been stored. In 2007 I tasted a delightful Wehlener Sonnenuhr Riesling Kabinett (S.A. Prüm), from the Mosel region in Germany and from the less than highly regarded 1987 vintage. At 20 years old, the wine still appeared lemon green, the colour having been preserved by the wine's high acidity, the result of having been produced in a cool year.

2.3.2 Rosé wines

Of all wine types, *rosé* is the category that is made to look attractive. Winemakers and marketing departments know that an appealing appearance is crucial to the drinker's perceptions of style and quality. The colour of rosé wines depends on several factors, particularly the production techniques, e.g. whether the colour is the result of skin contact in the press, or if the wine has been made by the *saignée* method, in which juice from a vat of fermenting, crushed red grapes is extracted after 8–24 hours or so. We may consider the colour of a rosé wine on the following scale:

pink — salmon — orange — onion skin

An orange tint can be a danger sign, indicating oxidation and that the wine is beginning to move to dryness which will be perceived as such on the palate.

2.3.3 Red wines

Red wines can vary in colour from purple to mahogany, loosing colour with age. Some grape varieties such as Cabernet Sauvignon and Syrah (also known as Shiraz) are deep coloured, especially when from a hot climate and from low-yielding vines; others such as Pinot Noir are relatively pale. Purple is indicative of a young red wine, some being so intensely coloured that they appear almost to be blue purple. As the wine begins to age the purple tones lighten to ruby, with further age to a warm brick red colour, and with full maturity possibly to garnet. Fig. 2.3 illustrates the appearance of young red Cabernet Sauvignon. Thus, we may consider the colour of a red wine on the following scale:

purple — ruby — garnet — tawny — brown

Red wines that have undergone lengthy barrel ageing, in which a controlled oxygenation has been taking place, lose colour faster than those bottled early and just aged reductively in bottle. A good illustration of this is the contrast between a 10 Year Old Tawny Port, which has been matured in pipe (cask), and a Vintage Port (*not* Late-Bottled Vintage-LBV) of about 12 years old, which has been aged reductively in bottle for all but the first 2 years or so. A wine that is brown in colour is tired, oxidised and probably undrinkable. Red and white wines that are totally oxidised are pretty much indistinguishable in colour.

2.3.4 Rim/core

The gradation of colour from the heart or core of the wine to the rim should be noted. Of course the greatest intensity of colour is at the heart, but in the area approaching the rim not only will the colour be paler but will also change. For example, a wine that is

Fig. 2.3 The appearance of a young Cabernet Sauvignon. To see a colour version of this figure, please see Plate 4 in the colour plate section that falls between pages 2 and 29

ruby coloured at its core may gradate to brick red or garnet or tones towards the rim, indicating maturity. As the rim of the wine touches the glass, the last millimetre or two will be water clear. The distance of the gradation in colour will vary from just a couple of millimetres in a young wine to perhaps a centimetre or more in a mature example. Mature white wines too will have considerable gradation in colour approaching the final few millimetres of the rim, which again will be water clear. The colour of the rim should be noted, and the width should be considered and noted on the following scale:

broad* — *medium* — *narrow

Fig. 2.4 Contrasting the colours and rims of young and mature reds. To see a colour version of this figure, please see Plate 5 in the colour plate section that falls between pages 2 and 29

Fig. 2.4 contrasts a young red Cabernet Sauvignon with a mature Brunello di Montalcino DOCG, some 23 years older, showing the striking difference in colour at the heart of the wines and the dramatic changes to the broad rim on the mature wine, paling to mahogany with orange tones as it laps the glass.

2.4   Other observations

2.4.1   Bubbles

Obviously, bubbles are a key feature of sparkling wines, but a small amount may also be observed on still wines.

Still wines

Occasionally, the presence of bubbles on a still wine could be indicative of a fault – either an alcoholic or malolactic fermentation is

Plate 1 Wine glass tilted to assess appearance

Plate 2 Looking down for an impression of intensity

Plate 3   The same wine as Plate 2 viewed tilted to 30°

Plate 4   The appearance of a young Cabernet Sauvignon

Plate 5   Contrasting the colours and rims of young and mature reds

Plate 6   The fine pearl-like bubbles of a good Champagne

Plate 7   Inconsistent bubbles in a charmat method sparkling wine

Plate 8   Wine showing thick, long legs

Plate 9 Part of the Molina vineyard in Curicó, Chile

Plate 10 Chablis Grand Cru vineyards of Preuses and Vaudésir

Plate 11 Inner staves inside an empty tank

Plate 12   Gravity feed winery at Rustenberg Estate, Stellenbosch

Plate 13   Gravity feed winery, showing press, at Rustenberg Estate, Stellenbosch

taking place or has taken place in the bottle (see Chapter 6). How-ever, still wines in good condition may contain bubbles. Gasses, namely carbon dioxide (CO_2), nitrogen (N) and argon (Ar) may be used as a blanket at various stages of winemaking in order to pre-vent oxidation or other spoilage. If a very fresh style of wine is being produced, it is common to flush the vats and blanket the wine with CO_2 at bottling time. Some of this gas may become dis-solved in the wine – this does not generally detract from quality and can often add a sensation of freshness. Some wines, e.g. wines from the Mosel region of Germany may naturally retain some CO_2 from the alcoholic fermentation. In the case of still wines, a brief observation of the size and quantity of bubbles should be made. The bubbles may just be on the glass, in which case they are likely to be large, on the rim, or in the heart of the wine.

Sparkling wines

The quality of the mousse is considered to be an essential part of the overall quality of sparkling wines. Notes should be made on the size, quantity and consistency of the bubbles. The bubbles may rise from the base of the cup or from a seemingly random point in the heart of the wine. Generally speaking, small bubbles are indica-tive of a desirable cool, slow second fermentation, especially one that has taken place in the bottle as in Champagne and other high-quality sparkling wines made by the **traditional method**. It should be noted that bubbles can vary somewhat according to the type and washing of the tasting glass (see Section 1.4). Fig. 2.5 shows the fine, continuous and even bubbles, with pearl-like strings, of a good quality Champagne, with a medium intensity of colour. Fig. 2.6 shows fewer and inconsistent bubbles in an inexpensive sparkling wine (a very pale Sauvignon Blanc) made by the **charmat** method.

2.4.2 Legs

One of the most misunderstood visual aspects of wine tasting is the presence or otherwise of legs. The wine should be swirled in the glass, held to eye level, waiting for several seconds and

Fig. 2.5 The fine pearl-like bubbles of a good Champagne. To see a colour version of this figure, please see Plate 6 in the colour plate section that falls between pages 2 and 29

viewed horizontally observed as to how the swirled wine runs back down the glass. If the liquid congeals into little tears, arches or rivers running down the glass these are called legs. The legs may be broad or narrow (thin), short or long, and run slowly or more quickly down the glass. Fig. 2.7 shows a wine with thick, long legs. Notes made should describe them as such, and subjective terms such as 'good legs' are best avoided. Wines that contain a

Fig. 2.6 Inconsistent bubbles in a charmat method sparkling wine. To see a colour version of this figure, please see Plate 7 in the colour plate section that falls between pages 28 and 29

high degree of alcohol, will normally show broad legs, formed by the difference in surface tension between water and alcohol and the differential evaporation of alcohol, influenced by sugar and glycol. Several authors and critics claim that legs are purely a sign of high alcohol or very high residual sugar, but this is refuted by examples of their presence in high-quality wines which are relatively

Fig. 2.7 Wine showing thick, long legs. To see a colour version of this figure, please see Plate 8 in the colour plate section that falls between pages 28 and 29

light in alcohol, for example, fine Riesling Kabinetts from the Mosel region in Germany. The amount of dry extract also contributes to legs. One of the most 'leggy' wines I have ever assessed was a dry white 1988 Château Miravel, Côtes de Provence AC, made from the Rolle variety, which yielded low at 30 hectolitres per hectare (hL/ha) and with an alcohol content of just 11.5%.

The amount and type of legs, if any, can be very dependent on the condition, and particularly the washing and drying of the tasting glass. I have poured the same wine into several apparently identical glasses, but washed at different times – each glass showed striking differences in the legs.

2.4.3 Deposits

Any deposit in the glass should be noted. These may comprise tannin sediments (which are simply the coagulation of phenolic substances) in the case of red wines or tartrate crystals in either reds or whites. Tartrate crystals are *not* a wine fault and are often visible in wines of the very highest quality. Thick deposits in red wines are usually tannin-stained tartrate deposits (especially in low pH wines). In white wines, however, they can look alarmingly like pieces of broken glass and really worry the consumer, but they are completely harmless.

The crystals are a precipitate of potassium bitartrate (most likely) or calcium tartrate (occasionally). They are very often found in bottles of German wine, which tend to be high in tartaric acid. They may precipitate if the wine gets very cool – perhaps in a cold cellar or refrigerator. Many winemakers go to great lengths to try to ensure that the crystals do not appear in the bottle. One step that can be taken is to chill the wine to $-4°C$ for 8–14 days immediately prior to bottling. This is known as cold stabilisation. The tartrates will then form and can be filtered out. An alternative method is to chill the wine to $0°C$ and seed it with fine tartarate crystals which act as nuclei to induce tartaric acid crystallisation and sedimentation. However, all these steps are expensive, and the acts of chilling to such low temperatures and fine filtration can actually reduce the quality of the resultant wine. The money and time spent on such treatments might be better invested in consumer wine education.

Nose

The olfactory epithelium, situated at the top of the nasal cavity, is a very sensitive organ. As will see in Chapter 4, the tongue reveals only a very limited number of tastes, and most of the 'taste' sensations are detected by the receptor cells of the olfactory epithelium, received either via the nasal or retro-nasal passage. The information is turned into electrical signals and sent via the olfactory bulb to the olfactory cortex in the brain. We will discuss the sensations transmitted via the retro-nasal passage in Chapter 4. There is no doubt that repeated and overexposure to particular smells reduces sensitivity to them, and this can be an issue for winemakers who are regularly and at times continuously exposed to odours such as sulfur dioxide.

When nosing a wine we are, of course, smelling the air space in the glass, above the surface of the liquid. It is important that there is plenty of head space for the aromas, the volatile compounds in the wine, to develop and the inwards tapering bowl of the tasting glass enables them to be retained therein. It is worth noting here that when pouring wines for general serving, the glasses should be filled to no more than 50% of their capacity in order for the nose to be appreciated. It is unfortunate that even in bars that offer an interesting range of wines by the glass, they are usually filled close to the brim to comply with legal sales measures. It is also sad that in the UK the minimum legal serving for a 'glass' of wine in a commercial establishment is 12.5 cL. In 2007 Selfridges was forced by trading standard's officers to desist from offering 'tasting size' pourings of some very fine wines into wine-tasting glasses.

The nose of the wine should be assessed in the following stages:

- Condition
- Intensity
- Development
- Aroma characteristics

As we will see below, the wine should be first nosed without swirling, then swirled around the glass to vaporise the volatile compounds and given a comprehensive nosing.

3.1 Condition

The initial nosing of the wine will assess the condition: many faults are apparent at this stage. The wine should not be swirled prior to this, and just one or two short sniffs are all that is required. Basically, we are checking if we want to go any further in the tasting ritual. For example, if the wine smells of damp sack or a damp, musty cellar, vinegar, struck matches or old cream sherry, it is faulty. Chapter 6 details faults and which of these are detectable upon the nose. Depending on the nature and severity of any fault revealed, a decision has to be made whether or not to proceed with the tasting of the wine concerned. A nose that is free from faults is described as clean.

A note should me made on the condition of nose:

clean − unclean + fault specified

3.2 Intensity

This is simply assessing how strong or 'loud' the nose of the wine is, and can give an indication of quality. A nose light in intensity may be expected from a simple, inexpensive wine, with a more pronounced nose indicative of higher quality. However, high-quality reds in particular can be very closed in youth. At the other end of the scale, there are some grape varieties that nearly always give a very intense nose especially aromatic whites including members of the Muscat family. We may consider and note intensity on the following scale:

light — *medium (-)* — *medium* — *medium (+)* — *pronounced*

3.3 Development

Development is assessing on the nose, the state of maturity of the wine. Wines-have a lifespan depending on many factors, and maturity should not be confused with age. Generally speaking, the higher the quality the wine, the longer is the lifespan. The finest quality reds, including many *cru classés* from Bordeaux, or classic wines from the northern Rhône Valley may require 10 years or more to approach their peak, whilst a simple Côtes du Rhône AC may be past its best at an age of 4 or 5 years.

To understand the concept of nose development, we need to consider the basic sources of wine aromas. These may be classified into three main groups.

3.3.1 Primary

Primary aromas are those from the grapes and are fruity or floral in character. These normally indicate a wine that is in a youthful stage of development.

3.3.2 Secondary

Secondary aromas are those from the fermentation – that is the difference between wine and grape juice. Numerous esters are generated in the fermentation and these are often assertive on the nose of young wines, imparting pear and banana characteristics. If present, the by-products of malolactic fermentation, battonage, and oak extracts are also included in secondary aromas.

3.3.3 Tertiary

Tertiary aromas are the result of the ageing process of the wine in tank, barrel and/or bottle. During this time there will be many chemical reactions. The wine will already contain some dissolved oxygen, and for wines that are barrel matured further oxygen may be absorbed through the cask. Thus, there will be a small amount of beneficial oxygenation taking place, increasing the aldehyde content of the wine. Wine matured in new barrels will also absorb oak compounds including vanillin, lignin and tannin; those matured in the second and third fill barrels will do so too, but to a lesser extent. Wines that are not matured in barrel may be micro-oxygenated and 'oaked' in other ways, and this is discussed in Chapter 9.

The maturation in bottle will result in changes to the volatile compounds of the wine. It is generally accepted that bottle maturation is of a reductive nature, that is, the changes take place anaerobically and the oxygen content of the wine is reduced. It is this process that makes bottle ageing a necessity for the maturation of many fine wines, especially full-bodied reds. A wine that does not have the capacity to reduce is a dead wine. However, *reductivity* is a fault and will be discussed in Chapter 6. As good wines mature the tertiary aromas develop, and these can exhibit a complex array of seamless, interwoven smell characteristics, at best all in total harmony. Tertiary aromas are those that often evoke the most descriptive comments such as saddles, Havana cigar, woodland floor, leather and autumnal gardens, and these observations may be noted later when we comment on the details of the wine's aroma. Many vegetal characteristics such as cabbage-like smells also belong to the tertiary group.

We may consider and note development of the nose on the following scale:

youthful — developing — fully developed — tired/past its best – deliberate oxidation?

It should be noted that some wines go from youthful to tired without ever passing through the fully developed stages. Examples include most Beaujolais AC and the inexpensive types of Valpolicella DOC.

But there are types of wine that are bottled fully developed, such as Sherry, Tawny Port and some sparkling wines.

3.4   Aroma characteristics

It should be pointed out at this point that many tasters and wine writers distinguish between the terms 'aroma' and 'bouquet'. The former refers to the nose which is classified above as primary aromas derived from the grapes. Bouquet encompasses the assembled nose characteristics resulting from changes that have taken place during fermentation (secondary aromas), and particularly maturation (classified as tertiary aromas). As we will see, the distinction is not clear-cut and I will use the word 'aroma' as an all-encompassing term.

Over four hundred wine odour compounds have been identified. Their concentrations vary from <100 ng up to 300 mg/L. The olfactory perception thresholds of the compounds vary from 50 pg to 100 mg/L. The compounds in grapes that are precursors of wine flavours include free amino acids, phospholipids, glycolipids, aldehydes and the phenols. Alkyl esters, a result of fermentation, are important compounds that give secondary aroma characteristics. Terpenes present in grapes are unchanged by the fermentation process and they contribute to primary aromas, although maturation and ageing of the wine may result in them being changed and contributing to tertiary aromas. Young wines made from grapes that have a high terpene content such as the Muscat family, Gewürztraminer and Riesling can have a nose that screams of primary fruit and show overt grape-like aromas. Many compounds that give varietal aromas remain largely unchanged by the fermentation process, and varietal aromas, e.g. the pronounced blackcurrant or cassis of Cabernet Sauvignon, are considered as primary aromas.

There has been considerable research over the last couple of decades into the compounds that contribute to wine aromas, particularly the varietal aromas. The aromas commonly found in wines made from Sauvignon and the compounds that give them are, for example, shown in Table 3.1.

Table 3.1 Thiols contributing to the varietal aromas of Sauvignon Blanc wines

Boxwood – broom	4-Mercapto-4-methylpentan-2-one
Citrus peel	4-Mercapto-4-methylpentan-2-ol
Grapefruit, passion fruit	3-Meracptohexanol
Boxwood – broom	3-Mercaptohexylacetate
Green pepper, grass	2-methoxy-3-isobutyl pyrazine (iBMP)

We may consider wine aromas in five basic groups:

fruit — floral — spice — vegetal — other

Each of these basic groups are composed of subgroups which in turn contain individual aromas and, when we taste the wine,

Table 3.2 WSET® Systematic Approach to Wine Tasting (diploma) aroma and flavour characteristics

Fruit	
Citrus	Grapefruit, lemon, lime
Green fruit	Apple (green/ripe), gooseberry, pear
Stone fruit	Apricot, peach
Red fruit	Raspberry, red cherry, plum, redcurrant, strawberry
Black fruit	Blackberry, black cherry, blackcurrant
Tropical fruit	Banana, kiwi, lychee, mango, melon, passion fruit, pineapple
Dried fruit	Fig, prune, raisin, sultana
Floral	
Blossom	Elderflower, orange
Flowers	Perfume, rose, violet
Spice	
Sweet	Cinnamon, cloves, ginger, nutmeg, vanilla
Pungent	Black/white pepper, liquorice, juniper
Vegetal	
Fresh	Asparagus, green bell pepper, mushroom
Cooked	Cabbage, tinned vegetables (asparagus, artichoke, pea, etc.), black olive
Herbaceous	Eucalyptus, grass, hay, mint, blackcurrant leaf, wet leaves
Kernel	Almond, coconut, hazelnut, walnut, chocolate, coffee
Oak	Cedar, medicinal, resinous, smoke, vanilla, tobacco
Other	
Animal	Leather, wet wool, meaty
Autolytic	Yeast, biscuit, bread, toast
Dairy	Butter, cheese, cream, yoghurt
Mineral	Earth, petrol, rubber, tar, stony/steely
Ripeness	Caramel, candy, honey, jam, marmalade, treacle, cooked, baked, stewed

flavours. Table 3.2 shows the aroma and flavour characteristics groups, subgroups and individual aromas/flavours.

All aromas sensed should be noted, and when detailing individual terms these may be linked to known varietal characters. For example, green apple, lime, peach and mango are just some of the aromas that may be associated with Riesling; strawberry, raspberry, red cherries, green leaf and mushroom are typical aromas associated with Pinot Noir. Any oak aromas including vanilla, toast, smoke, nuts and coconut should particularly be noted.

Palate

Palate is a convenient expression used to describe the taste of wine once it has entered the mouth. Under this heading we assess the taste and tactile sensations detected in the mouth and the flavour characteristics detected as a result of the wine's volatile compounds being breathed through the retro-nasal passage at the back of the mouth and transmitted to the olfactory bulb. It is important to breathe air through the wine as we taste to vaporise the volatile compounds, and a free passage is needed to and from the nose to enable transmission. If a person has a blocked nose, it is not just the sense of smell that disappears, but most of the sense of taste.

As we take wine into the mouth, taste, chew and dissect it before finally spitting the sensations develop. This evolution may be considered in stages: the initial **attack** as the wine is taken into the mouth, followed by **the development** on the palate where we perceive the flavour characteristics and intensity, and then the **finish** which comprises the final impressions of the wine including the balance. The **length** is a measurement of how long the sensations of the finish and aftertaste last. Tasters sometimes refer to the progressive sensations as 'front-palate', 'mid-palate' and 'back-palate'.

4.1 Sweetness/bitterness/acidity/saltiness/umami

The sensory cells contained within the 5000 or more taste buds on the tongue, although highly sensitive, can only detect four basic tastes: sweetness, bitterness, saltiness and acidity. These are the non-volatile compounds present in wine (although it should be remembered that acetic acid is volatile). Claims have been made of the existence of a fifth taste 'umami', the savoury taste of amino acids, but such claims are controversial. Of the four basic tastes, saltiness (comprising mainly sodium chloride) is not important in wine. The sensory cells on the tongue convert the detected tastes into electrical signals and send them to the brain's taste cortex. Traditionally, it has been accepted that different parts of the tongue detect these basic tastes. Many wine-tasting books and human biology texts illustrate a diagram of the tongue detailing these areas. However, as we saw in Chapter 1, this has been challenged by Linda Bartoshuk. For the purposes of this chapter we will rely on the traditional approach, as it is not disputed that the 'traditional' areas of detection do identify the tastes, only that the other areas do not. It is also accepted that the centre part of the tongue is much less sensitive to the basic tastes. There are tactile sensations of the wine that are also detected in the mouth, on the cheeks, teeth and gums, and these include tannin, body and alcohol.

When assessing the palate of a wine we consider the following headings:

- Dryness/Sweetness
- Acidity
- Tannin
- Alcohol level
- Body
- Flavour intensity
- Flavour characteristics
- Other observations
- Length

4.2   Dryness/sweetness

Before discussing perceptions of sweetness, we need to briefly consider grape sugars. Grapes contain glucose (grape sugar) and fructose (fruit sugar) which will be completely or partially converted to ethanol and carbon dioxide during the fermentation process by the action of yeasts. If there is insufficient natural sugar in grapes to produce a wine of the required alcoholic degree, in some countries the winemaker may add fructose to the must, a process generally known as chaptalisation – Jean Chaptal was a French Minister of Agriculture who first authorised the process in 1801 as a way of disposing of surplus sugar. In theory any added fructose should be fermented to dryness. Although extensively practised in many of the wine-producing member states of the European Union (EU), particularly in years of poor weather, there is discussion concerning the possible banning of the process following a proposal adopted by the European Commission in July 2007. An alternative to chaptalisation is the addition of concentrated grape must which of course contains glucose and fructose.

Sweetness, if any, in a wine can be detected on the tip of the tongue. It is important to remember than we cannot smell sweetness (sugar is not volatile) although the nose of some wines may lead us to expect that they will taste sweet. This may or may not be the case – for example, a wine made from one of the family of Muscat varieties may have a fragrant and aromatic nose reminiscent of sweet table grapes, but the wine, when tasted, may be bone dry. Other characteristics can also give an illusion of sweetness, in particular high alcohol levels (although too much alcohol can lead to a bitter taste) and vanillin oak. Thus, a high-alcohol wine that has undergone oak treatment can taste sweeter than the actual level of residual sugar. Pinching the nose whilst rolling the wine over the tip of the tongue can help the novice overcome any distortions that the nose may be giving. However, the acidity of the wine also impacts on the taster's perception of sweetness – the higher the acidity, the less sweet a wine containing residual sugar may appear to be.

Thresholds for detecting sweetness vary according to the individual: nearly 50% of tasters can detect sugar at a concentration of 1 g/L or less, with just 5% unable to detect sugars a less than 4 g/L.

Residual sweetness in a wine is due to fructose remaining after the fermentation. The level of residual sugar in white wine can range from 0.4 to 300 g/L. Most red wines are fermented to dryness or close to dryness, i.e. between 0.2 and 3 g of residual sugar. However, because dry wines are very fashionable, some wines are labelled or described as 'dry' when they are anything but. It is common for many New World branded Chardonnays to have between 5 and 10 g/L of sugar, and branded New World reds may contain up to 8 g/L, the sugar helping to soften any bitterness imparted by phenols. A little sweetness in a red wine can serve to balance any phenolic astringency. Wolf Blass, former owner of the famous Australian winery (now owned by a conglomerate), is quoted as saying: 'to sell your wine in Great Britain you must do two things: label it dry and make it medium!'

Whilst the sweetness in the world's greatest sweet white wines is grape fructose and glucose that remains after the fermentation has completed, there are other ways of producing sweet wines. The fermentation may be stopped by chilling the tank or by the addition of sulfur dioxide. To avoid further fermentation, the wine must then be filtered to remove yeasts and then bottled in sterile bottles. A method commonly used in Germany and practised in other countries is to blend a dry wine with up to 15% by volume of unfermented grape juice (*süssreserve*) followed by bottling. Such wines can often be identified by their overtly grapey character and lack of structure.

We may consider and note sweetness on the following scale:

dry — off-dry — medium-dry — medium — medium-sweet — sweet — luscious

4.3 Acidity

Acidity is particularly detected on the sides of the tongues and cheeks as a sharp, lively, tingling sensation. Medium and high levels of acidity encourage the mouth to salivate.

All wines contain acidity: whites generally more than reds, and those from cooler climates more than those from hotter regions. In the ripening process, as sugar levels increase, acidity levels fall and pH increases. Thus, a cool climate white wine might have a pH of 2.8, whilst in a hot climate red wine the pH might be as high as 4. Uniquely amongst fruits, grapes contain tartaric acid, and this is the main wine acid, although malic and citric are also important. These three acids account for over 90% of the acidity. Other acids present may include lactic, ascorbic, sorbic, succinic, gluconic and acetic acids. An excess of the volatile acetic acid is most undesirable. At very high levels it imparts a nose and taste of vinegar. If the grape must has insufficient acidity, the winemaker may be allowed to add acid, usually in the form of tartaric acid. Within the EU, such additions are only permitted in the warmer, southern regions.

Perception thresholds for acidity vary according to the individual. Nearly 50% of tasters are able to detect tartaric acid in concentrations of 0.1 g/L or less, and the remainder at between 0.1 and 0.2 g/L. However, sweetness negates the impact of acidity, and vice versa, and the balance between these is one of the considerations when considering *balance* as discussed below.

A wine's acidity may be assessed and described on the following scale:

low — *medium (-)* — *medium* — *medium (+)* — *high*

4.4 Tannin

Tannin is mostly detected by tactile sensations, particularly making the teeth and gums feel dry, furry and gritty. The sensations can be mouth puckering, and after tasting wines high in tannin, you want to run your tongue across the teeth to clean them. Hard, unripe tannins do also taste bitter. Tannin is a key component of the structure of classic red wines and gives 'grip' and solidity.

Tannins are polyphenols, the primary source in wine being the skins of the grapes. Stalks also contain tannins of a greener, harder nature and nowadays are, with some notable exceptions, generally excluded from the winemaking process. Oak is another source of tannin, and wines matured in new or young barrels, or otherwise oaked, will absorb tannin from the wood.

Tea contains tannin, and a good way of tasting the effect of various tannin levels without the influence of possible distorting factors such as acidity and alcohol is to make several cups of tea (no milk or sugar) with different amounts of maceration as shown below:

1 tea bag   30 seconds maceration
1 tea bag   1 minute maceration
1 tea bag   2 minutes maceration
2 tea bags   2 minutes maceration
3 tea bags   2 minutes maceration

Allow the tea to cool and then taste, ensuring that the liquid is thrown onto the teeth and gums. Note the increase in the gritty, drying sensations on the teeth and gums with each taste of stronger tea.

Tannin binds and precipitates protein. This, of course, is one of the reasons why, in general, red wines match red meats and cheeses successfully. This combination causes wines containing tannin to congeal into strings or chains as it combines with protein in the mouth, and thus our perception of tannin in a wine will change if we keep it in the mouth too long. To observe this, take a good mouthful of red wine low in tannin such as a Beaujolais AC, chew it for 20 seconds or so and then spit out into a white bowl. Now repeat the exercise with a red wine high in tannin such as a Barolo DOCG. It will be observed that the greater the tannin level in the wine, the more the wine will have formed these strings. Novice tasters often confuse the sensations of acidity and tannin. A classic Barolo DOCG, which is high in both, is a good example to taste to distinguish between them. The tannin gives the dry, astringent sensations on the teeth, gums and even hard palate. The acidity produces the tingling sensations on the sides of the tongue and cheeks.

It is often written that white wines contain no tannin. This is not true, although the levels are low compared with red wines. The grapes for white wines are pressed pre-fermentation, the solids are settled or the must otherwise clarified, and reasonably clear juice is fermented. Unless there is any period of skin contact post-crusher and pre-press, the phenolics in the skins will have limited impact. In the case of whole cluster pressing the role of phenolics is minimal. White wines that have been fermented or matured in oak barrels (or otherwise oaked) may contain considerable oak tannins.

The quantity of tannins in white wines ranges from 40 to 1300 mg/L, with an average of 360 mg/L. Red wines contain from 190 to 3900 mg/L, with an average of 2000 mg/L. Thus, it will be seen that whilst the average tannin level of red wines is six times that of whites, many white wines contain considerably more tannins than some reds. It is legal in most countries, including member states of the EU, for winemakers to add tannin, and this is often done to give some an over-soft red, a little more 'grip'.

Wine tannins should be assessed and noted for both the level and the nature. We may assess the level on the following scale:

low — medium (-) — medium — medium (+) — high

The nature of the tannins may be described as **ripe/soft** or conversely **unripe/green/stalky**. Texture too comes under this heading – tannins may be described as **coarse** that is rough, assertive and very gritty, or **fine grained** for those with a smooth, velvety texture.

4.5 Alcohol

Alcohol is detected on the palate as a warming sensation, on the back of the tongue, the cheeks and in the mouth generally. The weight of the wine in the mouth also increases with higher alcohol levels. Over-alcoholic wines will even give burning sensations.

The alcoholic range of light, i.e. unfortified, wines ranges from 7.5 to 16% abv (alcohol by volume). As grapes ripen, the levels of fructose and glucose increase, thus increasing the potential amount of alcohol. So we would expect wines from hotter climates to contain more alcohol than those from cooler regions. However, it is worth considering at this point that the amount of alcohol by volume in a wine will, as is the case with all other aspects of style, flavour and quality, depend on numerous factors:

- The **climate** of the region of origin
- The **weather** affecting the vintage in question
- The **grape variety** or varieties used
- The **soil** type and drainage

- **Viticultural practices** – including yield, training system, canopy management and choice of harvest time
- **Vinification techniques** – including fruit selection, yeast type, fermentation temperatures and whether or not an incomplete fermentation is stopped or the wine is blended with unfermented grape juice (*süssreserve*)

The average alcohol level of wines has increased during the last two decades. This is due to growers delaying harvesting until phenolic ripeness is reached (especially as the market now demands softer style of reds than in the past), the use of alcohol tolerant yeasts and the impact of global warming. In 1989, referring to Australian Cabernet Sauvignon, Bryce Rankine wrote in the reference work *Making Good Wine*: 'A ripeness of 10°–12° Baumé (18–21.6° Brix) is usual, which results in wine containing between about 10 and 12% alcohol by volume'. In 2008 any Australian Cabernet Sauvignon with less than 13% abv would be regarded as atypical – over-cropped and underripe. There are methods of removing alcohol from over-alcoholic wines including reverse osmosis and spinning cones, but these remain controversial.

We may assess the level of alcohol on the following scale:

low — medium (-) — medium — medium (+) — high

If a fortified wine is being assessed, the alcohol by volume will be in the range of 15–22%. We assess whether the wine has been fortified to a **low level** (15–16% abv), e.g. Fino Sherry or Muscat de Beaumes de Venise, a **medium level** (17–19% abv), e.g. Sherries other than Fino/Manazanilla, some *vin doux naturels* (VDNs) or a **high level** (20% abv or more), e.g. Ports.

4.6 Body

Body, sometimes referred to as weight or mouth feel, is more of a tactile than a taste sensation. It is a loose term to describe the lightness or fullness of the wine in the mouth. Body should not be confused with alcohol, although it is unlikely that a wine low in alcohol will be full bodied. Generally, wines from cooler climates tend to be lighter bodied than those from hotter areas. However, as with the other aspects of style, flavour and quality, the body of a wine will depend on the factors listed in Section 4.5. Particularly, certain grape varieties usually produce light-bodied wines, others full-bodied ones. Although it is a huge generalisation, wines made from Sauvignon Blanc or Riesling tend to be fairly light in body, whilst those made from Chardonnay or Viognier may be medium to full bodied. Of red grape varieties, Pinot Noir usually produces a lighter-bodied wine than Cabernet Sauvignon or Syrah.

Body may be assessed on the following scale:

light — medium (-) — medium — medium (+) — full

4.7   Flavour intensity

Flavour intensity should not be confused with body. A wine can be light bodied but with very pronounced intensity of flavour, for example, a fine Riesling from Germany's Mosel. However, as with the other aspects of style and quality, the flavour intensity of a wine will depend on the factors listed in Section 4.5. Of particular significance is the yield in the vineyard. The flavours of wines from high-yielding vines are generally more dilute and lack the concentration of those from vines with a low yield. Flavour intensity is one of the key considerations when assessing quality.

Flavour intensity may be considered on the following scale:

light — medium (-) — medium — medium (+) — pronounced

4.8   Flavour characteristics

As with the nose we may consider flavour characteristics on the palate in five basic groups:

fruit — floral — spice — vegetal — other

The reader is again referred to Table 3.2. Detailed notes should be made on the individual flavours perceived, and when noting

Table 4.1 Some white wine flavours and the grape varieties or other wine components commonly associated with them

Apple	Chardonnay (cool climate), Riesling
Apricot	Riesling, Viognier
Asparagus	Sauvignon Blanc
Banana	Chardonnay (hot climate)
Butter	Malolactic fermentation completed
'Catty'	Sauvignon Blanc
Citrus	Chardonnay (cool climate), Riesling
Coconut	Oak ageing
Cream	Malolactic fermentation completed
Creamy texture	Lees ageing
Elderflower	Sauvignon Blanc
Gooseberry	Sauvignon Blanc
Grapefruit	Chardonnay, Sémillon
'Herbaceous'	Sauvignon Blanc
Herbs	Pinot Grigio
Honey	Chenin Blanc, Riesling, Viognier
Kerosene	Riesling (aged)
Kiwi	Pinot Grigio, Sauvignon Blanc
Lanolin	Sémillon
Lemon	Chardonnay, Pinot Grigio
Lime	Riesling (moderate climate), Sauvignon Blanc
Lychee	Gewürztraminer
Mandarin	Sémillon
Mango	Chardonnay (hot climate)
Melon	Chardonnay (moderate climate)
Nectarine	Sémillon
Nuts	Chenin Blanc, Oak ageing
Passion fruit	Sauvignon Blanc
Peach	Chardonnay (moderate climate), Riesling, Chenin Blanc
Pear	Chardonnay (cool climate), Pinot Grigio
Pepper – bell (green)	Sauvignon Blanc
Petrol	Riesling (aged)
Pineapple	Chardonnay (hot climate)
Roses	Gewürztraminer
TCP	Noble rot (*Botrytis cinerea*)
Toast	Oak ageing
Vanilla	Oak ageing (especially American oak)
'Wet wool'	Chenin Blanc

individual terms these may be linked to known varietal characters. Table 4.1 lists some of the flavours that may be perceived in white wines and the grape varieties or other wine components commonly associated with them. Table 4.2 lists some of the flavours found in reds and their associated varieties and other wine components.

Table 4.2 Some red wine flavours and the grape varieties or other wine components commonly associated with them

Animal	Pinot Noir, high level of *Brettanomyces*
Aniseed	Malbec
Banana	Gamay, carbonic maceration
Blackberry	Grenache, Merlot, Shiraz (Syrah)
Blackcurrant	Cabernet Sauvignon
Bramble	Zinfandel
Cedar	Oak ageing, Cabernet Sauvignon
Cherry – black	Cabernet Sauvignon, Merlot, Pinot Noir (very ripe)
Cherry – red	Pinot Noir (fully ripe), Sangiovese, Tempranillo
Chocolate – dark	Cabernet Sauvignon, Shiraz
Cinnamon	Cabernet Sauvignon
Clove	Grenache
Coconut	Oak ageing
Game	Pinot Noir
Grass	Unripe grapes
Herb (mixed)	Grenache, Merlot, Sangiovese
Leafy	Pinot Noir
Leather	Shiraz (Syrah), aged wines
Liquorice	Grenache, Malbec, Cabernet Sauvignon
Meat	Pinotage
Metal	Cabernet Franc
Mint	Cabernet Sauvignon (especially cool climate)
Pencil shavings	Cabernet Sauvignon
Pepper – bell	Cabernet Sauvignon, Carmenère
Pepper – ground black	Shiraz (Syrah)
Pepper – white	Grenache
Plum – black	Merlot
Plum – red	Merlot, Pinot Noir (overripe)
Redcurrant	Barbera
Raspberry	Cabernet Franc, Grenache, Pinot Noir (ripe)
Roses	Nebbiolo
Smoke	Oak ageing
Soy sauce	Carmenère (very ripe)
Stalky	Wet vintage, unripe grapes, inclusion of stalks
Strawberry	Grenache, Pinot Noir (just ripe), Merlot, Tempranillo
Tar	Nebbiolo, Shiraz
Tea	Merlot
Toffee	Merlot
Toast	Oak ageing
Tobacco	Cabernet Sauvignon
Truffle	Nebbiolo

CHAPTER 4

4.9 Other observations

The *texture* of a wine should be considered, and the *balance* of the characteristics already discussed in this section. The easiest way to understand texture is by imagining running the tips of your fingers over the skin of various parts of the body of people of different ages and professions: the smooth, soft face of a model, the hands of a cashier, the weathered face of a fisherman, the pre-shave chin of a builder. The texture of a wine might be described in the range of:

silky — velvety — smooth — coarse

Bubbles or spritz, if present, give tactile sensations on the tongue. In a poor quality sparkling wine they are very aggressive, whilst a creamy feeling mousse is indicative of the well-integrated carbon dioxide in a good quality example.

Balance is the interrelationship between all the taste sensations and the components that create them. If any one or a small number of them dominate, or if there is a deficiency of any of them, the wine is unbalanced. An easy to understand example is that a white wine described as *sweet* or *luscious* but with a *low* acidity will be flabby and cloying – in other words unbalanced. A red wine with *light* body, *light* flavour intensity *medium (–)* alcohol but *high* tannin will feel very hard and astringent, again unbalanced. A balanced wine has all the sensations in proportions that make the wine a harmonious whole; in a well-balanced wine the sensations are seamlessly integrated.

If any component or a number of components are making the wine unbalanced, these should be noted. However, the state of maturity of a wine is an important consideration. Whilst a low-quality wine will never be in balance at any stage in its life cycle, high-quality wine, particularly reds, will often only achieve balance when approaching maturity. Tannins and acidity may dominate in youth, and the taster needs to evaluate all the components and the structure of the wine to anticipate how these will be interrelated at maturity. Balance is a major consideration when assessing wine quality.

4.10 Length

Put simply, 'length' is the best indicator of wine quality. Length is often referred to as aftertaste or finish although the expressions are not really synonymous. Crudely, length is the period of time that we can still taste a wine after we have swallowed it or spat it out. To be more precise, 'finish' refers to the final taste sensations of the wine as it is swallowed or spat. 'Aftertaste' encompasses the sensations that remain and develop as we breathe out, while 'length' is the measure of time for which finish and aftertaste last. To determine length, breathe out slowly, concentrate on the sensations observing any changes or development and count the seconds that the taste sensations last. The sensations delivered by poor quality and inexpensive wines will disappear after 5–10 seconds (short length), and any remaining sensations are likely to be unpleasant. Acceptable quality wines will have a length of 11–20 seconds (medium minus to medium length), good wine 20–30 seconds (medium plus to long length) and outstanding wines a length of 30 seconds or more (long length). Truly excellent wines may have a length that runs into minutes. It is important that throughout this test of length the sensations remain in tune with the actual taste of the wine, and also that everything remains in balance. The reader may wish to adjust the number of seconds timing given above for the various lengths to their own, individual perceptions.

Length may be considered on the following scale:

short — medium (-) — medium — medium (+) — long

If any unpleasant characteristics dominate the length, obviously they will have a negative effect on perceptions of quality and should be noted accordingly. For example, a wine with unripe tannins and other bitter compounds might have a medium or even long length, but the bitterness will dominate and the nature of the length becomes increasingly unpleasant.

Tasting Conclusions

Tasting conclusions bring together all of the information gathered during the assessment of appearance, nose and palate. Also added are the taster's judgements and value statements.

5.1 Quality

Quality judgements are framework dependent. This poses a dilemma. Do we consider the quality of a wine within the context of its peer group or against the entire wine world? Can a wine such as Beaujolais, which is made for early drinking in a soft, immediately approachable style, be described as outstanding quality even though it is carefully crafted, exquisitely perfumed, expressive of its origin and superior to most others of its type? The key to answering such questions is to be objective in the assessment and to note the quality of a wine, as perceived by the taster, according to quality and price levels within which the wine sits.

Quality may be considered on the following scale:

faulty — poor — acceptable — good — outstanding

5.2 Reasons for quality

It is, of course, very possible that many wines of outstanding quality are not to our palate, and many simple wines may, on occasions, be very appealing. The reasons for our quality judgements should be logical and as objective as possible. When reviewing the testing assessment, particular consideration should be given to the intensity and complexity of nose and flavours on the palate, and the balance and length. The following guidelines should form a framework for quality assessments:

- **Faulty** – showing one or more of the faults, detailed in Chapter 6, at a level that makes the wine totally unpalatable.
- **Poor** – a wine that is, and will always be, unbalanced and poorly structured. A wine with light intensity, very simple one-dimensional fruit flavours, maybe some flaws, and short length.
- **Acceptable** – straightforward wine with simple fruit, medium (-) or medium intensity, lacking in complexity, medium (-) or medium length.
- **Good** – an absence of faults or flaws, well-balanced, medium (+) or pronounced intensity and with smooth texture, complexity, layers of flavours, and development on palate, medium (+) or long length.
- **Outstanding** – intense fruit, perfect balance, very expressive and complex, classic typicity of its origin and very long length.

An outstanding quality wine will present the taster with an unbroken 'line', i.e. a continuity from the sensations of attack, when the wine first enters the mouth, through the mid-palate and on to the finish. It will develop and change in the glass and gain complexity; in other words, will not say all it has to say within a few seconds of the initial nose and taste. A really outstanding wine will also exude a clear, definable and individual personality, true to its origin, making a confident statement of time and place. It will excite in a way that seems to go beyond the organoleptical sensations – in other words, it will have the ability to move the taster in a similar way to a work of literature, art or music.

The 'line' of a wine, as detailed above, may be depicted visually in the form of a palate profile. This is a graph that illustrates the intensity and the texture of a wine from the attack (front palate) through to the mid-palate to the back palate and on to the finish. An example of a palate profile is shown in Fig. 5.1.

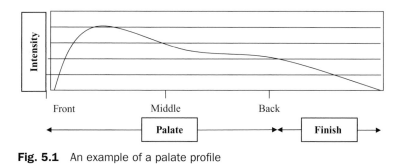

Fig. 5.1 An example of a palate profile

5.3 Readiness for drinking/potential for ageing

The topic of when wines are ready for drinking, and the assessment of this by tasting, is by nature complex. The life cycle of wines depends on several factors: colour, style, structure and particularly quality. Inexpensive wines are made to be drunk immediately, be they red, rosé or white. The reds will generally have been made without any lengthy post-fermentation skin maceration that would give firm tannic structure, and will have been fine filtered and stabilised before bottling. Three or four years in the bottle is the maximum keepability, and after this time they will have lost fruit and become 'dried out'. The further we move up the price scale, the more wines benefit from some bottle ageing, and fine red wines are designed for bottle maturation; the time taken for them to reach their peak and just how long they will remain there will vary according to the quality and style of the vintage as well as the origin of the wine. Intense fruit and the components of solid structure (high tannin, medium to high acidity and appropriate alcohol content) are the keys to a red that will improve in bottle. In youth these will be unknit and considerable bottle ageing will be required for them to evolve and integrate. High acidity is a great preservative in particular, but balance is crucial.

We may consider maturity on the following scale:

needs time (how long?) — ready to drink but can age (how long?) — at peak/drink soon — declining — tired/past its best

5.4 Price/value

The topic of blind tasting is discussed in Section 5.7. If the taster has not been informed of the price of the wine being assessed, it may be appropriate to make an estimate. Of course the conclusion as to quality is important, but this has to be related to the origin and maturity of the wine, and an accurate assessment of these is necessary in order to reach a reasoned estimate of price.

If the taster is aware of the price of the wine being assessed, it is relatively simple to relate the conclusions reached on quality to the price to decide whether the wine is poor, fair or good value for money. However, to some degree value judgements are also framework dependent, for some wines are by nature of their origin or scarcity more expensive. Of course at the level of great wines, it is probably not worthwhile attempting to make value judgements. For example, if the wines being assessed are Prestige Cuvée Champagnes, costing £70 or more, or *premier cru classé* Bordeaux, which may be priced in the hundreds of pounds, it is pertinent just to make a quality judgement. A poor wine is, of course, poor value however inexpensive it might be.

5.5 Identification/true to type?

When tasting blind the aim is to form an opinion on the characteristics of the wine being assessed. Depending on the circumstances and level of the tastings, factors to be determined may include **country/region/district of origin, grape variety/varieties, production and ageing methods** and age of the wine or vintage. The skill in making such judgements is based on considerable tasting experience together with a comprehensive knowledge of the factors that differentiate wines.

5.6 Grading wine – the award of points

The grading of wine is a controversial topic. There are those who claim that wines cannot be assessed by scoring points or on a star scale, and that the whole tasting process should be qualitative, not quantitative. Wine crosses the boundaries of art and science, and the exciting and complex characteristics of quality wines cannot be reduced to mere numbers. Countering this argument, many critics point out that in order to show which wines are simply superior to others they have to be rated on some sort of scale. Critics of musical performances or theatre often give star ratings, as do restaurant inspectors, reviewers of cars, washing machines and pretty much everything that is marketable. Of course, for the less knowledgeable consumer (or investor), knowing how well a wine has been scored makes the buying decision easier. Many consumers have consequently become bargain junkies, buying only the latest multibuy or BOGOF offer, while others have become points chasers.

Confusion is caused by the many different grading systems in use. Historically, a 0–7 scoring scale was often used, and of course star ratings, from 0 to 5 stars remain a popular system. To many critics, the use of scales such as these means that expression cannot be given to the small but perhaps significant differences in quality between wines in one of the grading bands. Accordingly, the preferred systems now mark the wines out of 20 (with steps that may or may not include decimal points), or out of 100, generally without decimal points. The marks can then, if deemed appropriate, be translated into a star rating. Both systems have strengths and weaknesses and we will examine them briefly. The scores given can be a sum of the individual scores for appearance, nose, palate, conclusions, etc., and in which case the assessor will mark within a framework or can simply be an overall score. Critics of the former method point out that simply adding the scores means that glaring weaknesses or flaws in one section only can mean that a wine is still scored reasonably highly.

5.6.1 Grading on a 20-point scale

This is the system used by judges at many wine shows, and also by several wine magazines, particularly those published outside of the Americas. The use of decimal points makes this a most expressive scale, as the 10–20 sections of the scale provide 100 grades.

Table 5.1 Scoring system as used by *Decanter* magazine

A score of 18.5–20 = outstanding (Decanter award) = 5 stars
A score of 16.5–18.49 = very good to excellent (highly recommended) = 4 stars
A score of 14.5–16.49 = good (recommended) = 3 stars
A score of 12.5–14.49 = fair = 2 stars
A score of 10.5–12.49 = poor = 1 star

The 20-point system used by the influential UK-published *Decanter* magazine is shown in Table 5.1.

The 20-point system used by *Wine* magazine, published in South Africa, is shown in Table 5.2.

It will be noted that these two systems are not entirely compatible.

5.6.2 Grading on a 100-point scale

This is the system used by the world's most influential wine critic, Robert Parker, by the US-published *Wine Spectator* and other magazines published in the USA. There are some crucial boundaries, particularly at 90 points, the level at which a wine is perceived as being outstanding. The influence of Robert Parker on buyers and producers of fine wines, particularly as far as Bordeaux is concerned, cannot be overstated. It may be said that if Parker gives a wine 89 points, the producer or merchant cannot sell it, and if he gives it 91, the consumer cannot buy it! Using the 100-point system, the lowest score for a poor wine is 50, and points are added for appearance (0–5), nose (0–15), palate (0–20) and conclusions, including quality and ageing potential (0–10). The system, as used in *Wine Spectator*, is shown as Table 5.3.

Table 5.2 Scoring system as used by *Wine* magazine (South Africa)

A score of 18–20 = top class, a masterpiece = 5 stars
A score of 16–17 = excellent, wine of distinction = 4 stars
A score of 15–15.5 = good to very good, fine character = 3 stars
A score of 14–14.5 = average, appealing = 2 stars
A score of 13 = acceptable, ordinary = 1 star

Extra half stars are awarded for wines that fall between the bands.

Table 5.3 Scoring system as used by *Wine Spectator* magazine (USA)

A score of 95–100 = classic, a great wine
A score of 90–94 = outstanding, a wine of superior character and style
A score of 80–89 = good to very good, a wine with special qualities
A score of 70–79 = average, a drinkable wine that may have minor flaws
A score of 60–69 = below average, drinkable but not recommended
A score of 50–59 = poor, undrinkable, not recommended

If scores are useful at all, it is only to people who understand the system. Much as low scores are obviously not used in marketing material, to the uninitiated scores of 15 out of 20, or 75 out of 100 may seem very good!

5.7 Blind tasting

5.7.1 Why taste blind?

Tasting wines 'blind' without the taster being given some or all the information about their identity is regarded as the most 'objective' way to assess wine. It is the method used in the tasting competitions discussed in Chapter 7, as critics and judges may be swayed by knowing which wine they are tasting, the reputation of the producer and their own previous perceptions of the wine. Blind tasting is also the best way of improving tasting technique, making the tasters rely on their own perception and apply their own descriptors. It is a valuable means to expanding the memory bank, particularly with regard to the relationship between the descriptors and the type of wine tasted. Depending on the objectives, the taster may have absolutely no information, or may be given certain relevant details, e.g. the wines are all Burgundy, are all made from one particular variety or are all in a certain price range – in other words, the wines are semi-specified. An alternative approach, sometimes called 'single blind', is when the details of the wines to be tasted are revealed beforehand, but not the order of tasting. Such frameworks can help concentrate the mind in evaluating wines for their quality, typicity, style and maturity.

5.7.2   Blind or sighted?

There is no doubt that knowing the identity of the wine to be tasted impacts on a taster's perceptions of it. There are, however, good reasons why the taster might wish to know the details of the wine being tasted. Firstly, the wine is immediately placed in context, including location within a region. Secondly, knowledge of the region and/or producer, e.g. their production methods and 'philosophy', can help the taster understand the wine. Thirdly, knowledge of a wine will help the taster place it at a particular point in its life cycle. Is it at it best? What is the potential to age based on reputation of the producer or region?

5.7.3   Tasting for quality

It can be argued that only by tasting blind can the taster come anywhere near to an objective assessment, as discussed in the Introduction to this book. This is especially important when, considering the factors that reveal quality, including complexity, balance and length, the taster is divorced from being influenced by preconceptions. If the objective of the tasting is to judge relative qualities, the wines chosen for the event should be comparable from this point of view, and in the broadest sense be stylistically similar – as we have seen, there is no point in trying to judge the quality of a Beaujolais against a *cru classé* Bordeaux.

5.7.4   Practicalities

When preparing a blind tasting, it is important to ensure that the wines really are presented blind. Unless the glasses are poured out of sight of the participants, the wines should be decanted into numbered neutral bottles. It is not sufficient merely to cover the bottles with foil or sleeves, for the shape and design of the bottle can sometimes reveal nearly as much information as the label. Numbered tasting mats as detailed in Chapter 1 are essential. Depending on the nature of the event, if there are more than 8 or

10 wines to be tasted, it is usually sensible to present them in 'flights', as this enables sharper contrasts to be drawn between the individual wines.

5.7.5   Examination tastings

If the tasting is for examination purposes, such as assessing a taster's abilities to accurately describe the wines and deduce the grape varieties, origins, vintages and quality, it is important that the examiner chooses representative samples. Each bottle of wine should be tasted by the examiner immediately prior to the examination and detailed notes made. If a wine is discovered to be unexpectedly out of condition, it may be included (provided that any duplicate bottles exhibit the same fault) unless the fault is so severe that it would seriously impact on the other wines tasted.

The taster should be cautious in approaching the wines. The order in which they are presented may not be a sensible order to nose or taste them, for example, a light delicate dry white will be almost impossible to taste after a rich sweet one, and a red with high and coarsely textured tannins will numb the palate for subsequent wines. Whilst sweetness and tannin cannot be seen or smelt, aspects of the appearance and nose will alert the taster to their probable presence. Thus each and every wine should be given an assessment of appearance before proceeding further. A gentle sniff of every wine too is essential before the taster decides in which order to undertake a detailed nosing, for a wine with pronounced intensity on the nose may numb the nose somewhat for a subsequent wine with light intensity.

The examination candidate should ensure that tasting notes are made against each of the tasting headings detailed in Chapters 2–5. Where appropriate the notes should be made in detail. Aroma and flavour characteristics in particular require a detailed analysis and description. However, verbosity should be avoided: the examiner should assess the candidates' ability accurately to describe the wine, not their skills as a wine writer. Each wine should be fully described individually, and not just compared and contrasted with the others. Vague expressions such as 'good', e.g. 'good acidity' and 'reasonable', e.g. 'reasonable tannins', should be avoided.

WINE QUALITY

When attempting to determine grape variety or varieties, origin, quality, and maturity, it is necessary to review all of the notes made and ensure that the conclusions are consistent. Whilst first impressions are often correct, a detailed deduction process is necessary in order to consider and review all possibilities. It is important that the tasting notes accurately reflect the wine assessed, and by jumping to a premature conclusion, the candidate may adjust the notes to fit. Students in professional examinations often erroneously believe that reaching the correct conclusion is paramount, whilst in fact it is upon a detailed and accurate description that the majority of marks are awarded. It is, of course, important that the conclusions are wholly consistent with the taster's descriptions and with each other. They must also be consistent with criteria that describe the wine that the taster has concluded it to be. By way of a very simple example, if the taster has concluded that a wine is blend of Cabernet Sauvignon and Syrah (Shiraz), it is totally incompatible to also conclude that it is an AC wine from Bordeaux, where Syrah is not a permitted variety. Finally, it is not unknown for students to erroneously conclude the identity of the wine as one that is outside of the syllabus or the framework of the examination question!

Wine Faults and Flaws

Whatever be the price point and whatever the quality level, aimed for wines are susceptible to faults. Many of the possible faults appear during the production process, and can be prevented or corrected at the appropriate time, but others may only manifest themselves when the wine is in the bottle and sometimes years after bottling. Some faults are so severe as to strip the wine of all semblance of quality or even drinkability.

Wine faults are either chemical or microbiological in nature. However, the issue as to whether an individual wine is faulty (showing one or more serious defects), flawed (showing minor defects) or sound is not necessarily straightforward. Apart from the matter of a taster's individual perception thresholds, there can also be dispute as to whether a particular characteristic is, or is not, a fault or flaw. For example, many Italian red wines have high levels of volatile acidity, which in wines from other countries might be considered flaws, but which contribute to the very 'Italian' character of the wines. *Brettanomyces* (Brett) is another controversial case and this is discussed below.

Detailed below are some of the most common faults and flaws encountered in bottled wine. The list is not exhaustive.

6.1   Chloroanisoles and bromoanisoles

Often described as 'cork taint' or 'corkiness', contamination of wine by haloanisoles, in particular chloroanisoles and bromoanisoles, is one of the most common faults and a topic that has been and continues to be the subject of considerable research. If the wine is heavily contaminated, the fault can be recognised instantly by nosing the wine – there is usually no need to taste. Upon raising the glass to the nose, the smell is instantly one of a damp cellar, a heavy mustiness, a smell of a wet sack, perhaps with tones of mushrooms or dry rot. If the wine is tasted, it will be dirty, fusty and earthy – rather like biting into a rotten apple. The fault is particularly noticeable on the back of the mouth. However, at low levels of contamination the symptoms detailed above may not be apparent, but the fruitiness of the wine is reduced and the taste is flat. Thus, contamination by haloanisoles in any amount is always a fault.

The main chloroanisoles and bromoanisoles involved in the contamination of wine are the following:

2,4,6-Trichloroanisole – perception threshold 1.5–3 ng/L
2,3,4,6-Tetrachloroanisole – perception threshold 10–15 ng/L, but
as low as 5 ng/L in sparkling wines
2,4,6-Tribromoanisole – perception threshold 3.4 ng/L
Pentachloroanisole – perception threshold >50 μg/L

Haloanisoles are transformed from the halophenols chlorophenol and bromophenol by the action of microorganisms, in particular filamentous fungi. One enzyme, chlorophenol-O-methyltransferase, is responsible for the transfer into haolanisoles. Chlorophenols are only found in nature due to anthropogenic origins. Chlorophenols have been widely used as cheap pesticides and fungicides during the past 60 years, and as they are not generally degraded by microorganisms, there is worldwide evidence of their pollutant effects. It is a sobering thought that chloroanisole contamination of wines did not exist before World War II – although Professor George Saintsbury refers to a corked wine in *Notes on a Cellar Book* published in 1920, we cannot know the precise fault noted. Banned in the European Union (EU) since 1991, chlorophenol usage continues in Africa, Asia and South America. Bromophenols are still permitted in the EU and are used as flame retardants and fungicides, and consequently 2,4,6-tribromophenol has been detected in many production cellars.

Cork bark may become contaminated with haloanisoles in the forest, either from the atmosphere or rainwater. Cork may also become contaminated in the production process or later. Thus, historically all related taints were described as 'corkiness' or 'corked'. However, wines from bottles sealed with closures other than natural cork can also exhibit the fault, thus exposing the myth that contaminated cork is the only cause. 2,4,6-Tribromoanisole taints are not derived from cork, but from within the winery. Sources of contamination by all the halonanisoles include oak barrels, bungs (especially those made of silicon), pallets, filters, plastics including shrink-wrap as commonly used on pallets of bottles, as well as bottling plants, winery structures and the atmosphere within wineries and cellars. There are several instances of entire cellar buildings having to be demolished and rebuilt due to contamination. The use of chlorine-based disinfectants and even tap water in the winery can also result in chloroanisole contamination. The haloanisoles are very volatile and migrate easily.

The major manufacturers of corks have undertaken considerable research and made substantial investments in recent years in order to try to eradicate cork contamination. Although this has resulted in a reduction in the problem, cork does remain one of the major sources of contamination. Of course, it is impossible for a wine producer to analyse every cork and a problem may only be discovered once a wine has spent years maturing in the bottle. An interesting point is that cork can act like blotting-paper and actually absorb 2,4,6-trichloroanisole from a contaminated wine, thus reducing the level of taint.

There are two other compounds not related to haloanisoles that can give musty taint to wines. These are 2-methoxy-3,5-dimethylpyrazine and 2-isopropyl-methoxypyrazine.

There has been much debate, at times very heated, about associated taints in recent years, the adversaries being the manufacturers of cork and the manufacturers of alternative bottle closures, and the pro- and anti-cork lobbies. There is divided opinion amongst trade buyers, writers and journalists. Some large grocery retailers now insist on synthetic closures or screw-caps when detailing the product specification with their supplier wineries. However, the use of synthetic closures is not always problem free as is discussed later in this chapter.

6.2 Fermentation in the bottle and bacterial spoilage

Fermentation in the bottle may be either a yeast fermentation or a malolactic fermentation. If a wine contains residual sugar together with live yeast cells, a re-fermentation in the bottle is possible, unless the alcoholic degree is above that at which yeasts will work, normally in the region of 16–17% abv. Further, if a wine has not undergone the malolactic fermentation prior to bottling and lactic acid bacteria are present, then this may take place in the bottle. The wine will contain bubbles of carbon dioxide, and even if these are not visibly detectable, the CO_2 may be felt as a prickle on the tongue. Active lactic acid can give a haze or silky sheen, particularly apparent when the wine is swirled around the glass. Cloudiness can be indicative of a fermentation in bottle or of bacterial spoilage. A mushroom or grey-coloured sediment is also sign of a (completed) bottle fermentation.

6.3 Protein haze

Protein haze is now a very rare fault. The wine will appear dull and oily, as a result of positively charged dissolved proteins being massed into light-dispersing particles. Protein hazes may be removed in the winemaking process by fining with bentonite.

6.4 Oxidation

Oxidation is a fault that is often apparent on appearance and certainly detectable on the nose of a wine. A white wine will look 'flat' – not at all bright, and in severe cases will deepen considerably in colour and start to look brown. A red wine will also look dull and take on brown tones. On the nose the wines will smell burnt, bitter, and have aromas of caramel or in severe cases the smell of an Oloroso Sherry. If the wine is tasted, it will be lacking in fruit, bitter, very dirty and short.

At some point in the life of the wine, it has absorbed so much oxygen that its structure has been damaged. This may have happened as a result of the grapes having been damaged, or having been subject to delays before processing, or due to careless handling of the wine in the winery including the failure to keep barrels topped up. However, the most common cause is when the wine in bottle has been stored badly or allowed to get too old. Bottles sealed with a natural cork should always be kept lying down to keep the cork moist and expanded in the neck. Wines sealed with plastic 'cork-shaped' stoppers are particularly prone to oxidation after some time in the bottle, as the stoppers harden and shrink. All wines have a finite life, and should not be allowed to get too old. Inexpensive wines, in particular, are usually made to be drunk almost as soon as they are bottled, and do not repay keeping.

6.5   Excessive volatile acidity

The total acidity of a wine is the combination of non-volatile or fixed acids, such as malic and tartaric acids, and the acids that can be separated by steam, the volatile acids. Generally thought of as acetic acid, volatile acid is composed of acetic acid and other acids such as carbonic acid (from carbon dioxide), sulfurous acid (from sulfur dioxide) and, to a lesser extent, butyric, formic, lactic and tartaric acids.

As the name suggests, volatile acidity is the wine acid that can be detected on the nose. All other acids are sensed on the palate. All wines contain some volatile acidity. If the level is low, it increases the complexity of the wine. If the level is too high, the wine may smell vinegary or of nail varnish remover (see also Section 6.12). On the palate the wine will exhibit a loss of fruit and be thin and sharp. The finish will be very harsh and acid, maybe even giving a burning sensation on the back of the mouth.

Acetic acid bacteria, such as those belonging to the genus *Acetobacter*, can multiply rapidly if winery hygiene is poor, thereby risking an increase in the volatile acidity of contaminated wines. The bacteria can develop in wine at any stage of the winemaking. The bacteria grow in the presence of air. If a red wine is fermented on the skins of the grapes in an open top vat, the grape skins will be pushed to the top of the vat by the carbon dioxide produced in the fermentation and will present an ideal environment for the growth of acetic acid bacteria. The bacteria can also be harboured in poorly cleaned winery equipment, especially old wooden barrels. Volatile acidity is also a by-product of the activity of *Brettanomyces* (see Section 6.8). Careful use of sulfur dioxide in the winery inhibits the growth of these bacteria, but overuse amounts to another wine fault as discussed below.

6.6 Excessive sulfur dioxide

Excessive sulfur dioxide can be detected on the nose – a wine may smell of a struck match or burning coke that may drown out much of the fruity nose of the product. A prickly sensation will often be felt at the back of the nose or in the throat. The taster may even be induced to sneeze.

Sulfur dioxide is the winemaker's universal antimicrobial agent and antioxidant. As stated above and below, its careful use inhibits the development of acetic acid and *Brettanomyces*. However excessive use, particularly prior to bottling, can lead to the unpleasant effects described above. It should be noted that the fermentation itself produces some sulfur dioxide. The amount of sulfur dioxide that can be contained in wines sold in the EU member states is strictly regulated by EU wine regulations. Higher levels are permitted for white wines than red, and the permitted level for sweet white wines is greater than that for dry whites. These are detailed in Table 6.1. It should be noted that some other countries permit

Table 6.1 Maximum permitted levels of sulfur dioxide within the EU

Dry red wine	160 mg/L
Dry white wine	210 mg/L
Red wine with 5 g/L sugar or more	210 mg/L
White wine with 5 g/L sugar or more	260 mg/L
Certain specified white wines, e.g. Spätlese	300 mg/L
Certain specified white wines, e.g. Auslese	350 mg/L
Certain specified white wines, e.g. Barsac	400 mg/L

higher levels; for example, USA and Japan both allow total sulfur dioxide levels of up to 300 mg/L, even in dry red wines.

6.7 Reductivity

Reductive faults comprise hydrogen sulfide, mercaptans and disulfides. All are recognised on the nose and all are the result of careless or uniformed winemaking. Hydrogen sulfide has a pronounced smell of rotten eggs or drains. The sensory threshold is in the region of 40 µg/L. Mercaptans can have an even more severe smell, where the odours are those of sweat, rotten cabbage or garlic or even skunk. The sensory threshold is in the region of 1.5 µg/L. Disulfides impart a smell of rubber or even burnt rubber.

Often, sulfur used as vineyard treatment is reduced to hydrogen sulfide in winemaking by the actions of the yeast. Wines made from grapes that are overripe and those grown on poor, nitrogen-deficient soils are prone to reductivity and some red grape varieties, especially Syrah (Shiraz), are particularly prone. The winemaker should take pains to keep sufficient levels of nitrogen (N) and oxygen (O) in the fermenting, must to counteract reduction. Diammonium phosphate may be added, usually at a rate of 200 mg/L of must. Its use is common in New World countries, but it does take a little colour out of the wine. Vigorous oxygenation during pump-overs is a proven method to counter the problem. This may involve splashing the wine drawn from the bottom of the fermentation tank into a container, pumping from this to the top of the tank and again splashing it over the cap of grape skins. In the event that a fermented wine exhibits the fault, chemical treatments (e.g. copper sulfate) may be used, but these do impact negatively on the fruitiness of the wine, and careless use can result in a copper haze. Another time-proven remedy is to place a piece of brass into a vat of wine that shows hydrogen sulfide.

Mercaptans are produced after the alcoholic fermentation by yeasts acting upon sulfur or hydrogen sulfide. Racking of red wines immediately after pressing can reduce the risk. In the case of whites wines prolonged lees contact, which can add delightful bread and yeast flavours together with a creamy texture, does present some risk. Mercaptans can also be suppressed by copper treatments, but disulfides cannot. They are the result of the oxidation and conversion of ethyl mercaptan. The group includes diethyl disulfide, dimethyl disulfide, dimethyl sulfide and ethyl sulfide.

A topic that has given rise to much research and considerable controversy during the last couple of years is that of post-bottling reductivity, particularly in relation to wines sealed with non-cork closures, especially screw-caps. Numerous incidences of this fault have added more fuel to the ongoing screw-cap versus cork debate.

6.8   Brettanomyces

Brettanomyces spp. (colloquially known as Brett) are yeasts that resemble *Saccharomyces cerevisiae*, but which are smaller. They are often present in the skins of grapes and cause a wine defect that is identified on the nose by a pronounced smell of cheese, wet horse, farmyards, baked orange or the plasters used for small wounds. The main compound responsible of imparting the farmyard or manure aroma is 4-ethylphenol. Other compounds are also present including 4-ethylguaiocol and isobutyric acid, the latter being responsible for imparting cheesy aromas. *Brettanomyces* contamination is often a result of careless winemaking and, particularly, poor hygiene management. They are very sensitive to sulfur dioxide and can be inhibited in winemaking by its careful use. Barrels can become impregnated with *Brettanomyces* and thus contaminate wine during maturation. Indeed, whole cellars can be contaminated, a situation that is very difficult to rectify.

The perception threshold for Brett is approximately 425–600 μg/L. Many producers in the Old World consider that a little Brett can add complexity to a wine and that it is very much an extension of the concept of *terroir*. Pascal Chatonnet, who has undertaken considerable research into contamination by haloanisoles and *Brettanomyces*, states that levels below 400 μg/L can add complexity, but as many as two-thirds of red wines contain levels higher than this. Many wines from the Rhône Valley in France show Brett, a well-documented example is the highly priced and generally very highly regarded Château de Beaucastel, Châteauneuf-du-Pape AC. Most New World producers consider any Brett to be a fault and expound that once Brett is allowed to remain in a cellar the impact on the wines is uncontrollable. That is not to say that there are not numerous examples of New World wines that exude considerable Brett character. Brett is most likely to taint wines that are high in alcohol and low in acidity, with excess nitrogen and with inadequate levels of free sulfur dioxide. There is no universal consensus on the 'correct' level of free sulfur dioxide in red wines, but 25–40 mg/L, depending on the acidity, is generally regarded as appropriate. Wines with a little residual sugar are particularly exposed. In recent years the demand for wines with mature, soft tannins has led to many producers waiting for full phenolic ripeness before harvesting. This results in high grape sugars (not always fermented out), high alcohol and pH levels which puts the wines at risk.

6.9 Dekkera

Dekkera is a sporulating form of Brettanomyces. It causes defects in wine, recognisable on the nose as a pronounced 'burnt sugar' smell or a 'mousey' off-flavour. It is often a consequence of poor hygiene in winemaking, especially in the barrel cellar. It can be inhibited by careful use of sulfur dioxide.

6.10 Geraniol

Geraniol is recognisable on the nose, as the wine smells like geraniums or lemon grass. It is a by-product of potassium sorbate, which is very occasionally used by winemakers as a preservative. It is important to use only the required quantity or the geraniol defect can arise.

6.11 Geosmin

Geosmin is a compound resulting from the metabolism of some moulds and bacteria including cyanobacteria, commonly known as blue-green algae. It is recognisable on the nose as an earthy smell of beetroot or turnip. Faults that can be attributed to geosmin may be caused by the contamination of barrels with the microorganisms that give rise to it or their growth on cork. However, geosmin can also be present on grape clusters, and if 2% or more of the clusters are contaminated, the resulting wine can show the defect.

6.12 Ethyl acetate

Ethyl acetate ($CH_3COOCH_2CH_3$) is recognisable on the nose as a smell of nail varnish remover or glue (both of which contain the compound). It occurs when ethanol reacts with acetic acid to produce ethyl acetate and water. The usual cause is a prolonged exposure of the wine to oxygen or the inadequate uses of sulfur dioxide during winemaking processes. It is present in low concentrations in all wines, and can be beneficial to aromas, especially in sweet wines. However, any excess, that is a concentration above the sensory threshold, is considered a major fault. The sensory threshold is approximately 200 mg/L, but this varies with the style of wine.

6.13 Excessive acetaldehyde

Acetaldehyde, formed by the oxidation of ethanol, is present in all wines in small amounts. In high amounts it imparts aromas similar to some deliberately oxidised wines such as *Vin Jaune*, and when in excess it gives very burnt aromas.

6.14 Candida acetaldehyde

This is a rare flaw or fault in bottled wine. The wine not only smells of straw with tones of fino-sherry, but also exudes a dirty character. A rogue yeast, *Candida vini*, is responsible for the defect. It is a film yeast that can form on the surface of wine in aerobic conditions and the root cause of the problem is usually a lack of care in keeping vats and barrels topped up.

6.15   Smoke taint

This is a problem found very occasionally in wines from Australia and South Africa as a consequence of bushfires or controlled burning taking place near vineyards. Affected wines may smell of burnt ash, smoked salmon or ashtrays. High concentrations of the compounds guaiacol and 4-methyl-guaiacol are found in tainted wines, but it should be pointed out that wines may contain low concentrations of these as a result of oak maturation or treatments. The sensory threshold is of approximately 6 μg/L, but bushfire-affected wine samples submitted to the Australian Wine Research Institute have shown guaiacol levels in excess of 70 μg/L. Of course, growers are aware if their grapes have been affected by smoke, but the fruit is usually harvested even if the wine made is considerably devalued. In Australia I have been shown and have tasted from vats containing 50 000 L of smoke-tainted wine. When asking the winemaker the destination for the product, it was suggested that a certain brand might benefit from the addition of some smoky, toasty flavours in the blend.

Quality – Assurances and Guarantees?

The quality of a wine may be assessed by undertaking a structured and detailed personal tasting, and we can have trust in our own judgement. The question now to be considered is: Are there third party assurances or guarantees of wine quality upon which we can rely?

7.1 Compliance with 'Quality Wine' legislation as an assurance of quality?

7.1.1 The European Union and third countries

Wherever wines are made, the product will to a greater or lesser extent be subject to compliance with local laws and regulations. In some countries this is largely a matter of ensuring that the product is safe to drink and correctly described on the label. In other countries, and particularly the member states of the European Union (EU), the relevant laws are more detailed and more restrictive. All wines produced within member states of the EU must comply with EU wine regulations regarding production, oenological practices, permitted additives and labelling. Council Regulation 1493/1999, as subsequently amended and detailed in Council regulations (particularly Commission Regulation 753/2002), lays down the permitted practices and required labelling information. It should be remembered that under EU regulations everything is prohibited unless specifically permitted. It is only subsequent to Regulation 1507/2006 that pieces of oak wood have been a permitted additive to wines produced in the EU, a practice long since legally practised elsewhere (and illegally practised within the EU).

Third country wines, i.e. those produced outside of the EU, must comply with the regulations if they are to be exported to the EU. Each wine-producing member state of the EU has its own wine laws which are subordinate to and comply with the requirements of the EU regulations.

7.1.2 Table wine and QWpsr

The EU categorises wine into two broad divisions:

- Table wines
- Quality wines produced in specific regions (QWpsr)

In most of the EU's important wine-producing countries, the broad *table wine* division is further divided into two categories:

- Table wine without a geographical indication
- Table wine with a geographical indication

The names that individual countries give to the table wine with a geographical indication category may or may not be well known: most drinkers will be aware of *Vin de Pays* (France), but few have heard of *Deutscher Landwein* (Germany). The rules of production are stricter than for the basic table wine (without a geographical indication) category and include a limitation upon yield in the vineyard. With the notable exceptions of eclectic producers who choose not to follow the restrictive rules for the 'higher' categories, most wines marketed as table wines (without a geographical indication) are, at best, undistinguished and at worst barely drinkable.

The concept of QWpsr is simple. Wines produced in a region, or more tightly defined area, should be typical and distinctive of the defined origin. The consumer does not want or expect a red Bordeaux to taste like a red Burgundy and we would expect a white wine from Mosel to have different characteristics to one from Baden. Each wine-producing member state that has an EU-recognised quality wine regime has its own name(s) for the QWpsr category or categories, together with detailed rules of production and labelling always within the already precise rules of the EU regulations. Do

Table 7.1 QWpsr categories in the most important wine producing countries in the European Union

France	Appellation d'Origine Contrôlée (AOC) or Appellation Contrôlée (AC)* Vin Délimité de Qualité Supérieure (VDQS)
Italy	Denominazione di Origine Controllata (DOC) Denominazione di Origine Controllata e Garantita (DOCG)
Spain	Denominacion de Origen (DO) Denominacion de Origen – Pago (DO Pago) Denominacion de Origen Calificada (DOCa)
Germany	Qualitätswein bestimmter Anbaugebiete (QbA) Qualitätswein garantieren Ursprungs (QgU) Prädikatsein
Portugal	Denominação de Origem (DO) Denominação de Origen Controlada (DOC) Indicação de Proveniência Regulamentada (IPR)

At the time of writing it is proposed to rename this as Appellation d'Origine Protégée (AOP)

these rules guarantee that a wine designated and labelled as a quality wine is just that? Indeed in the case of Italy which, as with the other important EU-producing countries, divides its QWpsr into two broad categories, the 'highest' level is called 'Denominazione di Origine Controllata e Garantita', certainly implying guarantee of quality. Table 7.1 lists the names given to the QWpsr categories in the most important wine-producing countries in the EU.

Without doubt, it is the French term 'Appellation Contrôlée' (AC) that is the best known to wine lovers and, with apologies to other member states, we will limit our discussion here to France. At the time of writing it is proposed to rename the term as 'Appellation d'Origine Protégée (AOP)'.

7.1.3   The concept of Appellation Contrôlée

The term 'Appellation Contrôlée' (often referred to in full as 'Appellation d'Origine Contrôlée' – initials 'AOC') encompasses 99% of QWpsr production in France. The remaining 1% of QWpsr production is of wines categorised as Vins Délimités de Qualité Superieure, historically an important category, now only used as a staging post before a particular wine area is deemed to have achieved the necessary standards for full AC status and, at the time of writing,

Table 7.2 Factors controlled by AC regulations

Delimitation of vineyard area
Permitted grape varieties
The maximum yield per hectare
Methods of viticulture
Methods of vinification
The minimum alcoholic degree of the wine
The wine must pass a tasting test
The wine must pass a laboratory analysis

scheduled to be phased out. Appellation Contrôlée is (in theory at least) a guarantee of origin and a very basic typicality. It should be noted that AC does not only apply to wines, but also to many French foods, e.g. AC Roquefort (cheese). The system is administered by the Institut National de l'Origine et de la Qualité.

Factors controlled by AC regulations are detailed in Table.7.2.

Essential to the concept of AC is the delimitation of the area of production. The area can be the following:

- **A region,** e.g. Bourgogne AC (Burgundy)
- **A district** within a region, e.g. Chablis AC
- **A group of villages** e.g. Côte de Beaune Villages AC
- **A commune** (Parish), e.g. Beaune AC
- **An individual 'Premier Cru' vineyard,** e.g. Beaune Toussaints Premier Cru AC
- **An individual 'Grand Cru' vineyard,** e.g. Corton Charlemagne Grand Cru AC

The theory is the more precise the appellation, the more individual the wine should be. It is where the grapes are grown that determines the AC, not where the wine is made, e.g. wine made from grapes grown in Chablis will bear the AC Chablis, even if the wine was made elsewhere in Burgundy, e.g. the Côte d'Or (although the wine must be made within the region, a criterion that does not apply to all QWpsr).

As stated, AC is not a guarantee of quality, but the more precise the AC, the better the land is officially rated for its potential for higher quality production and the more strict are the legal criteria that apply, which *may* result in better quality. Much depends on the producers and their commitment to quality. One of the drawbacks of the system is that individual appellations have a commodity value. For example, in Chablis a wine from an individual Grand Cru vineyard usually has a value of double or treble a wine that just bears the district appellation, Chablis AC, however good the latter might be. This of course does little to encourage producers of the 'lower' appellations to make the very best wine possible from their land.

As we have seen, it is not only the vineyard area that is delimited for AC wines. There are many other factors that will contribute to taste, style and individuality, and these are embodied in the AC laws. One of the key factors is, of course, the grape variety or grape varieties from which the wine is made. In some regions or districts, the AC laws restrict the possibilities to just one variety (e.g. all Chablis AC and nearly all other white Burgundy must be made from Chardonnay), while in other areas the grower has a wider choice and blends the wine from a number of different varieties. However, one of the key points is that the varieties permitted for AC wines are those that are traditional to the area. If growers wish to plant a non-traditional variety, they will generally not be able to sell the wine with the name of an AC although it is possible that a Vin de Pays category is available to them.

The density of vine planting may be specified. For example, in the Bordeaux region, vines planted in the communal Pauillac or St Estèphe appellations must be planted at minimum density of 6500 vines per hectare. Those for the district appellations of Médoc or Graves, a minimum of 5000 vines per hectare is required, and under the latest changes those for the regional Bordeaux Supérieur or Bordeaux ACs, 4500 and 4000 vines per hectare, respectively.

The tasting and laboratory analytical tests for Appellation Contrôlée have historically, in many regions including Bordeaux, not been of the finished wines. The wines have been assessed part way through their development. In other words, the wines tasted have never been marketed as such, for after the tasting they will have been subject to further ageing, in vat or barrel, blending and preparation for bottling, including perhaps fining, filtering and cold stabilisation. It can be argued that such a tasting test does little to guarantee that the purchaser will receive even a typical and acceptable wine, let alone a quality product. However, as from 2008 changes in the legislation require the wines to be assessed at bottling time. The vast majority of wines are accepted, and it is rare that a wine is declassified or rejected. There is, of course, an appeals process for those that are.

It is worth noting here that producers in third countries sometimes borrow the names and labelling terms of perceived illustrious appellations, although such labelled wines may not be imported into the EU. Fortunately, after pressure from the EU, the use in Australia and USA of terms such as 'burgundy' and 'chablis' has now ceased but 'port' lingers, especially in South Africa. The terms 'premier cru' and 'grand cru' legally protected and with particular significance in France mean nothing in the New World: in South Africa there are several wines marketed under the label 'premier grand cru'. At the time of writing they sell for 19 rand (approximately £1.35) per bottle – one is described in the *John Platter South African Wine Guide* as 'mildly fruity'.

7.2   Tasting competitions as an assessment of quality?

The assessment of the quality of various wines by a columnist for the purpose of conveying the results and giving the consumer a 'best buy list' is a regular feature of weekly newspaper columns, monthly magazine articles and annual wine guides. Some of these opinion formers are expert tasters and well educated in vinous matters, others sadly are not. As we saw in Chapter 5, there are also tastings undertaken by panels of expert tasters for wine magazines and annual tasting competitions such as the *International Wine Challenge* and the *International Wine and Spirit Competition*. The results are categorised by point tables or the award of medals. Such tastings are usually carried out under blind conditions (although this is not the case for the well-known *John Platter South African Wine Guide*) and can be a useful source of information for the prospective purchaser, but there are several caveats:

(1)   Especially at the large competitions, where tasting panellists assess large numbers of wines, it can be the biggest, loudest wines that win through. It is sometimes said of Australian wine that the more the medals that appear on the bottle, the less likely you are to want to drink the whole bottle.

(2)   The wines tasted will have been subject to recent movement and some might not be shown in the best of condition.

(3)   The competitions require entry fees from the producers/agents and for this and other reasons many decide not to enter.

(4)   And perhaps most importantly, the assessment is merely a snapshot of the particular bottles tasted at that time. Large producers may keep many of their wines in vat until such time as they are needed for the market. Bottling is expensive and bottled stock is space consuming. Further, many wines may be kept 'fresher' in vat – this is especially true of aromatic white varieties such as Sauvignon Blanc. Accordingly, there will be several bottlings taking place throughout the year of 'medal winning' wines. At very least, each wine will be adjusted and stabilised for bottling, but vats may have developed differently and blending may take place. So, the wine being shipped may not be identical to the medal-winning wine.

On this last point it is interesting to note that Article 21 of EU regulation 753/2002 states:

> For the purposes of the third indent of Annex VII(B)(1)(b) to Regulation (EC) No 1493/1999, awards and medals may be featured on the labels of table wines with a geographical indication and quality wines psr provided that these have been awarded to the batch of wine concerned in a competition authorised by a Member State or third country and run with complete impartiality. The Member States and third countries shall notify to the Commission the list of authorised competitions.

With regard to this regulation, in my experience most wines labelled as medal winners are done so illegally.

Finally, in this section it is worth noting that critics and judges may have strong views on how wines should taste, as exemplified by the dissenting views of two influential wine critics on one particular wine: 2003 Château Pavie, Saint-Émilion Premier Grand Cru Classé. Jancis Robinson MW described the wine as 'ridiculous wine, more reminiscent of a late-harvest Zinfandel than a red Bordeaux' and scored it 12/20. Robert Parker described it as 'a wine of sublime richness, minerality, delineation and nobleness', and scored it 98/100.

7.3 Classifications as an official assessment of quality?

Classifications, such as the Saint-Émilion classification mentioned above, provide an official assessment of the quality of a property's wine. They are most important in the Bordeaux region. The most famous classification is that of 1855 which classified the red wines of the Médoc district (together with Château Haut Brion situated in the Graves district) and the sweet white wines of Sauternes. This classification was drawn up by wine brokers, based on the prices that the wines achieved and, with the exception of one well-known promotion (Château Mouton-Rothschild in 1973) and one not at all well-known addition to the list (Château Cantemerle in 1861), has not been amended. Some other districts in Bordeaux have since been subject to classification, using many assessment criteria as a basis, including tasting of the wines. The wines of the Graves district were last classified in 1959, and those of Saint-Émilion (which in theory re-classifies every 10 years) in 2006. However, the 2006 Saint-Émilion classification was suspended by a court in March 2007 following complaints from four producers who disputed the result. The classification was subsequently reinstated in November 2007 by France's highest administrative court, the Conseil d'Etat, but incredibly declared invalid by a Bordeaux court on 1 July 2008. The relevance and validity of all the classifications is always subject to debate, especially as owners come and go. Land may be traded within the boundaries of the Appellation and quality may rise or fall. Interestingly, the superiority of the 1855 First Growths – Château Lafite-Rothschild, Château Latour, Château Margaux, Château Haut-Brion and Château Mouton-Rothschild (added in 1973) (red wine classification) and Château d'Yquem (sweet white wine classification) – is today not really disputed. High prices mean more money for investing in quality, which means higher prices. Those who question the relevance of classifications also point out that the two properties in Bordeaux that now normally achieve the very highest prices, Le Pin and Château Pétrus, lie in the district of Pomerol, which has never been classified. The rise in reputation of Le Pin is staggering – the first year of production under the name was 1979, the year that the property was purchased by the Thienpont family. The 1982 vintage now commands a price in the region of £2000 per bottle.

7.4   ISO 9001 certification as an assurance of quality?

ISO 9000 is an identifying code for a series of quality standards published by the International Organisation for Standardisation (ISO), based in Geneva. Central to the series is the standard, ISO 9001:2000. This is a quality system standard that serves as a model for the development, implementation and maintenance of quality management systems. ISO states that the ISO 9000 series is concerned primarily with what an organisation does to fulfil:

– 'the customer's quality requirements, and
– applicable regulatory requirements, while aiming to
– enhance customer satisfaction, and
– achieve continual improvement of its performance in pursuit of these objectives.'

Quality management systems developed against ISO 9001:2000 may be used by wine producers in the effective control of the production and commercial processes that operate to satisfy customers' needs. Effectively, the quality management system defines the way in which customers' requirements will be fulfilled and quality improvement achieved. ISO 9001:2000 requires the implementation of a documented quality management system based on a quality manual, management and operating procedures and quality records. The use of third party auditors, or certification bodies, enables the objective and independent assessment of compliance with the international standard and, hence, worldwide recognition of capability and competence in quality management. ISO 9001 has received criticism from some quarters because of the apparent focus on the achievement of customer satisfaction through the prevention of nonconformity at all production stages. The idea exists that with ISO 9001 rules must not be broken, even if by doing so a superior product might be achieved. It is believed that with ISO 9001 quality is seen as something that fulfils the customers' requirements, not their aspirations. It is thought that quality is dumbed down and, consequently, many top-quality producers reject ISO 9001 and the concepts it stands for. It is considered that using ISO 9001 is not a guarantee of the quality of the finished product, yet there are certified producers that unquestionably rely on the standard and believe that by doing so they produce 'quality' products however ill-defined the word might be. Whilst visiting a large bodega (winery) in Argentina, I asked the winemaker about steps they might take to lift the quality of the product. The reply was somewhat frightening: 'our customers can be assured our wines are amongst the best. We have ISO 9001 and HACCP!'

7.5 Established brands as a guarantee of quality?

Consumers' faith in large brands is built around their perceptions that the brands they choose give them a guarantee of quality and consistency and that the product or service is one that addresses their individual wants and needs. With regard to wine, the successful brands have a large market share and a seemingly ubiquitous presence on the supermarket shelves. However, it is rare to find a wine writer with a kind word to say about the quality and individuality that some of these offer, particularly the super-brands. Many brands have ranges of product pitched at different price levels. An example is *Penfolds* (owned by Fosters FMEA) which markets wines that sell from as little as £5 a bottle and up to £100 a bottle or more (Penfolds Grange). It should be recognised that many companies that sell branded wines produce them in more than one country.

It is not easy to establish the point at which a wine from a long-established producer makes the transition to branded product in the modern sense of the term. For instance, is the well-known and respected label Château Margaux a brand? Certainly the well-known Champagne houses are now established as brands and the perceptions of quality associated with the Champagne marques are high. But Champagne itself is a relatively small production region (presently 33 000 ha, although this is set to increase) and the price commanded for the wine means that high standards in the vineyards and cellars are the norm. Grape prices are high too: in the 2006 harvest the price paid by the houses was €4–5 per kg (€4000–5000 per tonne or £2800–3500 at the exchange rate at that time). The reader may wish to compare this with the prices achieved in the 2006 harvest in South Africa, as detailed in Chapter 9.

It is interesting that the rise and fall of brands depends as much if not more on marketing budgets and brand-building activities than any variation in the quality of the product. The biggest selling brand of South African wine in the UK is *Kumala*. At the time of writing the brand owner is *Constellation Brands* (the largest wine production company in the world), which had purchased *Vincor* who in turn had purchased *Western Wines*, the original brand owner. It is pertinent to note that until 2007 *Kumala* owned neither a vineyard nor a winery, and the wines were all sourced by one broker. For some time after the acquisition by *Constellation Brands*, the development of *Kumala* seriously faltered, resulting in a sales decline in the UK from over 3 million cases per year to less than 2 million. Mostly as a result of this, sales of South African wine in the UK declined by 11%. In any event brands, almost without exception, have a finite life. Those drinkers who in the 1970s enjoyed(?) drinking Crown of Crowns, Hirondelle and Don Cortez have long since moved on.

In the last 20 years, New World wine brands have had far more impact in the UK market than those of the Old World. At the time of writing, nine of the top ten brands for UK off-sales are New World wines. The undisputed largest fine wine region in the world, Bordeaux, with 123 000 ha of vineyard, more than the whole of South Africa, and an average annual production of approximately 800 million bottles, has only one brand (Calvet) in the top 50 (at the number 50 position, and Calvet's wines from Burgundy helped make up this ranking). To many consumers, the New Word generally provides consistent, easy to understand wines, very much fruit driven and designed to be instantly approachable.

So do brands offer the guarantee of consistency and quality referred to in the first paragraph of this section? In order to invest in brand building and maintenance, the owner needs to be reasonably sure of the continuity of the supply of grapes. The concept of blending often has negative connotations amongst serious wine lovers, even though nearly all wines are blended to some extent. However, blending lies at the very heart of creating and maintaining the style of wines for the big brands. Fruit may be sourced from the producer's own vineyards, growers under contract, or in the spot market. Wines may be made in one or more wineries, and blenders have many components to maintain the style. However, blending large quantities for consistency, by definition, means that blending cannot be for the best possible quality, for some blends will not meet the hard-to-reach standards. Blending for consistency means that the individuality of a particular vintage, region or even winemaker (for they come and go) has to be negated. Wine is an agricultural product and the style and quality of fruit from any region or district will vary according to the weather patterns of the year. Individual properties may decide not to make their 'top' wines in poor years, and the brands that usually source fruit from those districts will face particular challenges. Of course by striving to be consistent, there may need to be a good deal of technical adjustment to the wines – additions of acid, tannin, etc. Non-branded wines may well have such adjustments, with the purpose of making the particular wine as good as the winemaker can achieve, rather than to maintain the consistency the brand owners require.

Finally, to counter the 'strength of brands is their consistency' argument, the consumer is not guaranteed a consistent product anyway. Supermarkets in particular often buy product from more than one source. Brands may vary from lot to lot and particularly from market to market. Some of the best-known brands of Champagne produce to a different specification according to the market to which the product is destined. Champagne buyers in Britain generally prefer their Champagnes with much more autolytic ageing on the lees (the legal minimum is 15 months) and with a lesser dosage than required by the buyer in Germany. So if the supermarket bypasses the regular source of supply and buys via alternative sources, it is possible to find very different cuvees standing side by side on the shelves.

7.6   Price as an indication of quality?

It can be argued that the market is the final arbiter of wine quality. If the product is superior, it will command a higher price than lesser wines. Even before they were bottled in 2007, the finest *premier cru classé* Bordeaux wines of the excellent 2005 vintage were being traded at a bottle price of £450–600 ($900–1200). However, as already stated, markets are subjects of hype and fashion. It is interesting to note that in the late nineteenth century it was the wines of Germany that commanded the highest prices, more than any *cru classé* Bordeaux. For example, the 1896 wine list of the highly reputed London merchant Berry Bros and Rudd offered:

1862 Rüdesheimer Hinterhaus   *200 shillings (£10) per dozen*
1870 Château Lafite   *144 shillings (£7.20) per dozen*

The Lafite was the most expensive Bordeaux on offer and from the great 1870 vintage! Even as recently as 1963, Peter Dominic were offering:

1959 Château Gruaud Larose   *18 shillings/6d (95P) per bottle*
1959 Wehlener Sonnenuhr
  Spätlese (J. Bergweiler)   *32 shillings (£1.60) per bottle*

1959 was an excellent year in both Bordeaux and Germany.
   In 2007 Berry Bros and Rudd offered:

2000 Château Gruaud Larose   *£89 per bottle*
2001 Wehlener Sonnenuhr Spätlese
  (Markus Molitor)   *£19.50 per bottle*

It is argued that the price of any wine can be no more than the market will bear. Whilst producers need to recoup production costs and make a profit, this does not mean that the price of the wines necessarily reflects differences in the production costs of individual products. As we have seen, many brands comprise a range of levels, and often all wines in the level are pitched at the same, or broadly similar, price whatever the production cost. The price will be determined with regard to the perceived market, and the image to be maintained. Appellations too have a 'commodity value', which reflects their image: the producer of the very finest Muscadet can only dream of achieving the price achieved by the maker of a mediocre Sancerre.

Quality – The Natural Factors and a Sense of Place

Wines may be considered as being produced in three differing conceptual styles:

- Variety driven
- Fruit driven
- Site driven

Varieties, to many consumers, are like brands. They offer a statement of style and taste, and by providing a wine that adheres to the benchmark characteristics of the variety of which it is labelled, the expectations are satisfied. As one moves up the quality scale, the more expressive of the variety the wine should be. Fruit-driven wines exude masses of primary fruit enhanced, in good wines, by the secondary and tertiary characteristics. Site-driven wines may be more restrained, yet more individual, and express their origin, in all its complexity, rather than a more simplistic statement of variety or fruit.

There are other conceptual styles by which wines may be classified, for example, those which are:

- **expressive of raw materials** – grape variety, region;
- **made to meet a market** – big-volume brands which sometimes use consumer panels to choose style;
- **winemaker concept wines** – often blends of grape varieties or districts – e.g. Penfold's Grange, many Bordeaux Cru Classés and 'Super Tuscans'.

8.1   Typicity and regionality

Until 40 years ago the origin of wines was generally regarded as the main factor that determined their style and to a large extent their quality. Historically, ripeness or otherwise was accepted as in the hands of the gods, according to the luck of the year. Classic wines such as *cru classé* Bordeaux were fabled for their *goût de terroir* and were made to be laid down for a generation or more – indeed the great but firmly structured years such as 1870 and 1928 did not reach maturity for 50 years. Each region made wines in the way that they had been made for countless years. Grape varieties were little discussed – they were much less understood than today and the varieties grown were those traditional to the area. Chardonnay (usually referred to as Pinot Chardonnay) and Pinot Blanc were often confused as the 'white Burgundy noble grape'. By 1968 Viognier was a variety on the verge of extinction, with just 14 ha remaining in France at Condrieu and Château Grillet. The first ever varietal Chardonnay was produced in Australia in 1971 by Murray Tyrrell in the Hunter Valley. New World countries were of small importance to the UK market.

Today, the *cru classés* are more successful than ever, but it is unlikely that any of the richer, less tannic wines they produce will last for 50 years. Grape varieties and brands rule the wine roost. Australia, having overtaken France, is at the time of writing the number one supplier to the UK off-sales market by volume and value, although following droughts resulting in reduced harvest quantities this situation may not last. The vast majority of Australian wines on sale in the UK bear no more precise a regional origin than that of the 'super-zone' South Eastern Australia, which encompasses all the wine-producing regions of New South Wales, Victoria and South Australia and comprises approximately 95% of Australia's production. Grapes and bulk wine may be transported many hundreds of miles, giving the large producers ample components to make their multi-regional blends. The fruit is ripe and the wines correctly made. Within this vast ocean of fruit and variety-driven wines, there are many that are good and some that are very good. But, by definition they lack a sense of place. Australia has, of course, many regions that have that sense of place. The most southerly region in

South Australia is Coonawarra, relatively cool, and at its heart a fa-mous strip of *terra rossa* soil. Well-made, classic wines from Coon-awarra tell you that this is where they are from. Cabernet Sauvignon particularly is distinctive, exuding purity, and expression, savoury, minty earthy, dusty and edgy. The Semillons from Margaret River in Western Australia exude a grassiness that is as distinctive as it is exhilarating.

8.2 The impact of climate on quality wine production

The climate of the wine region has a considerable impact on the style and quality of the grapes produced and thus the wine made. In cooler regions grapes may not fully ripen, and these regions can also be subject to a considerable annual variation in weather patterns, thus producing good or poorer quality harvests. It should also be noted that the quantity produced may vary substantially. Generally speaking, cooler climates will produce grapes with lower levels of sugars and higher levels of acidity than those grown in hotter regions. With not enough sun and heat, red grapes have weak concentration, green tannins and an acidity that can be swingeing. Excess humidity can encourage the most unwelcome onset of powdery mildew, downy mildew, and botrytis in the form of grey or bunch rot. If there is insufficient grape sugar, this may be compensated for (in regions where such adjustment is legally allowed) by the addition of sugar, or concentrated must in order to make a wine of the required alcoholic degree. Similarly de-acidification, using Acidex® or alternatively calcium carbonate, may be required (in regions where such adjustment is legally allowed).

Hot regions too can present climatic problems to the producer. Grapes may ripen very quickly, giving high sugar levels (and high pH) but without having had sufficient time for flavour development, or conversely with extreme heat the flavours getting burnt out. Grapes can be affected by sunburn – one way to help counter this is by leaving a denser leaf canopy on the western side of the vines to shade the grapes from the hotter, afternoon sun. If temperatures exceed 38°C, sugar accumulation in grapes more or less stops because the vine is using all of its energy for staying alive. Provided the vine has enough water, photosynthesis will continue, but the vine's metabolism speeds up with higher temperature faster than the rate of increase of photosynthesis.

If the temperatures exceed 43°C, there is a risk of the vines dying. In the Peel region, close to Perth in Western Australia, temperatures peaked on Boxing Day 2007 at 45°C, devastating both the quantity and quality of the 2008 harvest.

Producers in regions that are situated on the viticultural edge where, given a warm summer and fine autumn, there is enough sun to slowly ripen the grapes with a lengthy hang time on the vine claim that these conditions not only maximise flavours but give naturally balanced fruit. Similar statements are made by those in climates with a large diurnal temperature range – the cool nights lock in colour, acidity and flavour. It should be noted that large diurnal temperature ranges are not restricted to continental climates: Napa, for example, has a very maritime climate with relatively small differences between summer and winter temperatures, but high diurnal ranges. The individual **mesoclimate,** the local climate within a particular vineyard or part of it (which is influenced by many factors, including altitude, aspect and the proximity of water and trees), impacts on the quality and style of fruit produced. The *Grands Crus* vineyards of the Côte d'Or in Burgundy are situated in various parts (depending partially on aspect) of the middle of the narrow slope – near the top of the slope it is too cool and windy, and near the bottom the soils are less well drained.

8.3 The role of soils

The physical, chemical and biological properties of the soil, both top- and subsoil, impact on style and quality of the finished wines. The most important physical characteristics of different soil types are those that govern water supply, water holding and drainage – put simply, quality wine is not produced from poorly drained vineyards. The texture of the soil will depend on the proportions of gravel, sand, silt and clay and affects the vine's ability to take up water, nutrients and minerals. Texture is very hard to adjust. Soil structure results from the type of aggregates (lumps of soil particles) and the spaces between them, and can be altered by preparation and vineyard management techniques, such as the addition of gypsum (calcium sulfate, $CaSO_4 \cdot 2H_2O$). Compaction, often caused by the passing of tractors and other machinery, must be avoided as the soil will be deprived of oxygen and the ability of the roots to grow will be restricted.

Before planting a new vineyard, a full soil analysis should take place which examines the chemical and biological composition and balance. The pH of soils suitable for viticulture lies between 5.0 and an absolute maximum of 8.5. The pH of the soil has an impact on the style and quality of wine produced. All other factors being equal, vines grown on a high acid (low pH) soil will produce grapes with a lower acidity than those grown on a low acid (high pH) soil. It is soil ion availability that aids or impedes acid retention in the grapes. Regular additions and adjustments to the soil, e.g. the addition of lime to acidic soils, may be needed when the vineyard is established. Of course, it is important that the vine variety, clone and rootstock planted should be compatible with the soil (and climate).

8.4 Terroir

The French word 'terroir' defies simple translation. It encompasses the type and quality of soil of a vineyard site, together with the mesoclimate of the area and topographical factors. Mankind too has modified terroir by a history of viticultural practices including reshaping vineyards such as making (and removing) terraces, the installation of drainage systems and changing the chemical and biological structure of the soil by the addition of chemicals and organic matter.

The soil type and drainage are key components in the concept. Historically, it was generally believed that poor (low in nitrogen) well-drained soils were best for vine growing. However, recent research in the Bordeaux region indicates that a relatively high nitrogen content will increase the aromas of aromatic varieties such as Sauvignon Blanc. Certain soil types are regarded in some countries as inferior to others. For example, in France both sand and alluvial soils are generally thought not to be conducive to producing high-quality wine, although producers in parts of New Zealand would strongly disagree. Other soil types are generally favoured, including gravel, granite, clay and limestone and particularly the *argilo calcaire* mix of the last two.

Professor Gerard Seguin of the University of Bordeaux has studied the chemical properties of soils around Bordeaux, including the top Grands Crus in the Médoc, and the lesser properties. The research reveals that the top châteaux have soils with a high percentage of acidic gravel and pebbles, some 50–62% as compared with only 35–45% in the lesser crus. The top châteaux have soils that are naturally poor in nutrients and particularly the first metre of depth has a modified chemical profile. The soils are deficient in magnesium due to high levels of potassium. Seguin considers that this imbalance with the resultant ion antagonism effect lowers vine vigour and yields, and that the finest wines, even within the same property, come from such parcels. The soils are also high in phosphoric acid and organic matter. The clay contents of the soils have been raised, and the sand content decreased. All these modification can be attributed to the owners being able to afford over many decades manure, compost and other amendments which improved soil structure, permeability and resistance to erosion.

The physical properties of the soil are crucial to making great terroir, particularly the drainage and the amount of water stored in the soil at the depth to which roots penetrate. The best soils are free draining and the water supply to the vines is regulated. The roots of the vine penetrate deep and assimilate the minerals and trace elements found at greater depths. Good drainage is particularly important after a period of heavy rainfall as waterlogged soils result in the vine taking water via the surface roots and also encourage excessive vigour and attacks of mildew. Seguin's research into the Grands Crus Classés of the Haut-Médoc reveals that the level of their water table drops as the growing season progresses. From the time of veraison in August, when the grapes change colour and ripening commences, the water table is too low to supply the vines. Thus during the crucial late summer ripening period, unless there is rainfall, the energy of the plant goes into ripening the fruit rather than producing vegetation.

It is only in the last decade or so that producers in the New World have really begun to discuss their terroirs. The concept was, and it must be said still is by many, regarded as an Old World marketing gimmick and something that could be used as an excuse for poor winemaking. To the new generation New World producer, armed with the latest research from UC Davis or Waite, many of the expounded terroir characteristics of Old World wines were attributable to poor oenological practices or contaminations: minerality was down to excessive sulfur dioxide and leather and horsey tones was nothing more than *Brettanomyces*. A few favoured areas in the New World were extolled as being individually superior, a good example being Coonawarra in South Australia, with its famed *terra rossa* strip of soil. Incidentally, this area was the subject of a very protracted boundary dispute, for 'Coonawarra' on a label gave a marketing advantage.

However, there are three reasons that the terroir concept is less important in many New World regions:

(1) Irrigation – although this is essential in areas of insufficient natural rainfall, as we have seen above, expressions of terroir are reduced when the vine has an easy supply of surface water. This does not mean that irrigation is a bad thing, but it needs careful control. We will look at the concept of inducing moderate hydric stress in the vines in Chapter 10.

(2) At many sites there is only a generation or two of winemaking experience, contrasted with up to 2 000 years in the classic regions of Europe. There has been considerable growth of the amount of land planted for wine grape production in the last 20 years in the key New World producing countries. For example, in Australia the amount of land in production in 1990 was 42 000 ha and in 2007 it was 174 000 ha.

(3) Many New World vineyards are massive. For example, the Molina vineyard of Viña San Pedro in Chile's Curicó Valley comprises 1200 contiguous hectares. In Burgundy the seven delimited, named, Chablis Grand Crus vineyards, discussed below, total 100 ha. Part of the Molina vineyards is shown in Fig. 8.1, and the Chablis Grand Cru vineyards of Preuses and Vaudésir in Fig. 8.2. When large vineyards in the New World are machine harvested, the different terroirs, if they exist, are not isolated and the grapes have no chance of separate vinification.

Fig. 8.1 Part of the Molina vineyard in Curicó, Chile. To see a colour version of this figure, please see Plate 9 in the colour plate section that falls between pages 28 and 29

Fig. 8.2 Chablis Grand Cru vineyards of Preuses and Vaudésir. To see a colour version of this figure, please see Plate 10 in the colour plate section that falls between pages 28 and 29

There are many New World producers who take terroir to heart. Philip Woollaston of Woollaston Estate in Nelson, New Zealand, says: 'I believe the place is the most important influence on quality and style. A wine is defined by place, not by variety and not by brand. A brand is nothing. Place is defined, and cannot change. Brands come and go.' In the nearby region of Marlborough, Saint Clair Estate produces a range of wines, individually vinified, bottled and marketed by the number of the vineyard block. The contrast between these wines made using identical techniques is remarkable. On a larger scale, New Zealand's largest producer Pernod Ricard produces and markets a range of wines defined by individual Marlborough terroirs.

There are many in the wine industry who challenge the concept of terroir. Mike Paul, former chairman of *Wine Intelligence*, believes that the New World wine revolution of the 1990s saw terroir diminish in importance and that 'Australia has proved that the winemaker is the hero, and terroir is just one ingredient in the mix'. However, a tasting of several wines from the same vintage, made from grapes of the same variety, clone and vine age, grown using identical viticultural techniques and produced in one cellar using identical winemaking techniques, reveals that terroir has a profound influence on the flavours of the finished wine. For example, in the district of Chablis, the seven named Grand Cru vineyards lie contiguously in a hillside arc just north of the village. The elevation and aspect vary somewhat and the soils, basically fossil-rich Kimmeridgian clay, also have subtle differences. The wine from each of the vineyards tastes different, even when the vinification is identical. Fig. 8.3 shows the named Chablis Grand Crus vineyards and indicates the differences in the flavours of wines produced therefrom.

8.5   The vintage factor

Each vintage in each region will vary in **quantity** and **style** as well as **quality**. Cold weather early in the season, particularly frost at budding time and in the weeks that follow or windy weather at flowering time, can result in a considerably reduced crop. Hail in late summer can devastate a crop and even the vines. Such problems

---

Chablis Grands Crus

The Chablis Grand Cru vineyards cover a total area of 100 ha (the size of just one of the larger Médoc châteaux) and comprises seven named AC Grand Cru Vineyards:

Vineyard name	Size and styles of wine
Bougros	12.6 ha. The vineyard is situated at the western end of the arc of the Grands Crus and produces wines with a vibrant but solid, masculine character
Preuses	11.4 ha. Producing wines with a succulent almost exotic, ripeness, fleshy, richer and fatter than the other Grands Crus
Vaudésir	14.7 ha. Ripe wines with an underlying mineral and spicy character, feminine and very complex
Grenouilles	9.3 ha. Softer, delicacte wines that are very fruity, racy and elegant
Valmur	13.2 ha. The wines show delicate floral aromas and are perhaps the softest of the Grands Crus
Les Clos	26 ha. The largest of the Grands Crus, and considered by many as the finest, producing wines that are rich and powerful yet steely, with exhilarating minerality
Blanchot	12.7 ha. The wines show floral and mineral tones and are of delicate structure. They can be firm when young, but develop wonderful complexity

Fig. 8.3 Chablis Grands Crus

are not exclusive to Europe. Many New World regions are at risk, e.g. Mendoza, Argentina's largest wine region, can be victim to cold, frost, wind and hail. Plantings on slopes are generally at less risk of frost as the dense, cold air flows down the slopes. Growers may use a number of methods when attempting to protect their vines, including oil burners, spraying water on the vines via an aspersion system and windmills. Getting the air circulating is effective. In New Zealand, which has had an increasing problem with frosts since 2001, helicopters are often used to achieve this in addition to the methods above. It is not uncommon for there to be over 100 helicopters in the air over the region of Marlborough.

Drought can also result in reduced crop, as happened in Europe (especially much of France) in 2003. Even with established irrigation systems in place, New World countries can suffer too. This was the case with the 2007 harvest in Australia where the lack of rainfall caused restrictions in water use. Of course, the reduction in yield by some weather factors may result in an increase in quality, but the key is that the limited fruit production must be clean and balanced.

While plenty of summer sunshine is clearly desirable, if it gets too hot – above 38°C – the vines' metabolism can shut down, resulting in underripe fruit. A little rain in the spring or early summer is generally desirable, for if the vines are too heat stressed, quality will also be reduced. Perhaps most desirable of all is a period of dry, warm sunny weather in the weeks prior to harvesting. Many otherwise great vintages have been wrecked by autumn rains arriving while the fruit is still on the vine.

Vintages vary in style from the big, rich wines produced in years when it has been hot to the lighter but very good wines produced in years with less sunshine. Grape varieties ripen at different times; for example, Merlot ripens up to 3 weeks before Cabernet Sauvignon. Thus, in Bordeaux for instance, if the rains come in late September, the Merlots will probably already be fermenting and wines dominated by this variety may be excellent, whilst the Cabernet Sauvignon dominated wines will be poorer. On the other hand, if August and early September have been cool, with excessive rain, but then there is an Indian summer later in the month and into October, it is the wines majoring on Cabernet Sauvignon that may excel.

Rain at harvest time can be a nightmare, creating a mood of depression throughout a wine region. *Bortytis cinerea* as grey rot can become established on grapes very quickly. The fungus thrives in a humid atmosphere, causing infected grapes to expand by taking up water from the soil and resulting in those in the middle of the bunch becoming compacted and damaged. Bunch thinning can help to reduce this problem. If fruit is delivered to the winery wet, it will dilute the juice. Mechanical harvesters can prove their worth here, for if rain is forecasted they can bring in the crop very quickly. However, many small properties may rely on the use of a contract machine, and this will not be available to them at the time it is most needed. If a short period of rain is forecasted, provided the fruit is healthy, a grower may take a gamble and delay picking until afterwards, hoping for an Indian summer to achieve more ripeness.

If hand harvesting takes place, pickers may be instructed to leave damaged or rotten fruit on the ground. Sorting, as detailed in Chapter 10, is essential in wet harvests.

The problem of dilution can be addressed by the employment of **must concentrator** machines in the winery – these cost up to £90 000, so they are not for the small producer. Vacuum evaporation of excess water results in the concentration of sugars. Machines employing reverse osmosis, a technique only authorised in the European Union since 1999, concentrate sugars and flavours. The largest must concentrator machines can extract 1500 L of rainwater per hour. It is interesting that the Appellation Controlée authorities in France do not consider that the use of must concentrators negates the concept of *terroir*, but when two Bordeaux properties covered their vineyards with plastic sheeting during the 2000 growing season, the resulting wines were not accepted as AC (resulting in the world's most expensive Vins de Table).

The vintage factor adds a sense of time to that of place. The variation in style, detailed flavours and quality, reflecting the growing year, adds another layer of individuality to the wines of a particular region or producer. This is unlike every other drink, where style and consistency from batch to batch is the norm. A vertical tasting, that is tasting a number of vintages of the 'same' wine (albeit at different stages of maturity), is invariably one of the most fascinating and revealing of wine-tasting experiences because of the variation in product characteristics that such a tasting reveals.

Constraints upon Quality Wine Production

'Mother nature isn't always kind' is the oft-heard lament of those working with vines and their produce, basically an agricultural product turned by art and science into the most talked of beverages, 'but it's much kinder than the government and the bank'. In other words, it is the business of producing wine that can put more constraints on quality than natural forces.

9.1 Financial

The number one aim of any wine producer, grower or winemaker is to make a profit. Indeed most producers constantly strive to maximise profits. This may be achieved by the following:

(1) Increasing prices – this may involve 'repositioning' the product. The price charged cannot, of course, be more than the market will bear.
(2) Increasing quality – this may boost sales or lead to (1) but will usually also involve an increase in costs, although it can often be achieved simply by improving practices.
(3) Increasing production – this may result in reduced quality, always assuming the market will take such an increase with no downward pressure on price.
(4) Increasing the market's perceptions of the product which can lead to (1).
(5) Reducing costs – this may result in reduced quality as discussed below.

9.1.1 Financial constraints upon the grower

A grower, who is not also a wine producer, will sell grapes either to a cooperative of which he or she is a member, or directly or indirectly to a wine-producing company. The price paid will depend on many factors. These include contractual agreements, the state of the market, the variety grown and the quality and sanitary condition of the crop.

Growers may have long-term contracts with wine producers, often requiring them to adhere to stipulated practices and to produce a crop to the buyer's specification. It is not unknown for buyers to renege on such contracts in times of financial difficulty or glut, and if the grower is tempted to seek legal address, they will probably be bankrupt by the time the case is heard. However, without such contracts the grower is at the mercy of the market, and this can prove disastrous. Cooperative members too are exposed when market prices fall. The years 2004–2006 saw growing production surpluses in many countries, including Australia and South Africa, with resultant fall in grape prices. Table 9.1 shows the decline in average price achieved per tonne for sales by members to cooperative cellars in South Africa for the three most important red varieties – Cabernet Sauvignon, Merlot and Shiraz – in the harvests from 2001 to 2007.

The 2006 and 2007 figures cited represent a figure below production cost. In 2007, the price of 1.1 kg of Shiraz, the approximate amount needed for a 75 cL bottle of wine, was a little under 2 rand (approximately 13 pence at the then rate of exchange). It

Table 9.1 Average prices of grapes (rand per tonne) sold for production of wine by members to cooperatives in South Africa during the period 2001–2007

Variety	2001	2002	2003	2004	2005	2006	2007
Cabernet Sauvignon	4849	4590	4080	3228	2386	1925	1537
Merlot	4300	3987	3602	2477	1997	1731	1585
Shiraz	4159	3717	3524	2633	2091	1854	1562

Figures supplied by SAWIS.

Table 9.2 Average prices of grapes (rand per tonne) sold for wine production by growers to private producers in South Africa during the period 2001–2007

Variety	2001	2002	2003	2004	2005	2006	2007
Cabernet Sauvignon	5423	5458	5605	5476	4212	3428	3026
Merlot	5285	5331	5477	5327	3963	3186	2960
Shiraz	5106	5111	5294	5036	3907	3027	3255

Figures supplied by SAWIS.

must be stressed that these were average prices, and growers of top-quality grapes could sell at a premium. The prices achieved by growers selling their grapes to private producers were significantly higher, although it still showed a decline in the latter part of the period. Table 9.2 shows average price achieved per tonne for sale by South African growers to private cellars for these grape varieties during the years 2001–2007.

It should be pointed out that fashionable white varieties such as Sauvignon Blanc did register an overall price increase during this period.

In times of financial pressure, savings have to be made. Labour may be cut back, and soil adjustments and pest and disease control measures may be limited. If total income is down, the temptation is to increase yield in order to have more product to sell, thus to maintain cash flow. Such yield increases further reduce quality (and result in a further reduction in prices obtained).

9.1.2 Financial constraints upon the winemaker

The relentless pressure on producers from retailers striving to maintain price points, offer promotions and improve margins means that quality compromises often have to be made in the winery. The market rules and costs must be made to fit. Fruit selection may be minimal or non-existent, as witnessed at some large wineries by the 10 tonnes tipper trucks discharging into the receiving hoppers that feed the crushers. At every stage the must and wine will be speeded on its progress through the winery, for time is money. This will involve clarification and stabilisation techniques that can compromise quality – e.g. tangential filtration and flash pasteurisation. One key area where costs can be contained is in the 'oaking' of the wine.

Oak treatments as a substitute for barrel ageing

The cost of barrel maturing wines is not just the capital costs of the barrels, but the increased amount of labour required and the loss of a small amount of wine due to evaporation. The high cost of barrel maturation is prohibitive for wines with inexpensive price points, so numerous alternative ways of oaking wines have been developed. These include oak powder, granules, chips, rods, ladders and staves. The rate of usage and approximate cost per litre of wine for the various products is shown in Table 9.3.

Table 9.3 Rate of usage and approximate cost per litre of wine for various oak products

Product	Typical dimensions	Usage	Unit price	Cost per litre of wine
Large tank staves	2200 × 100 × 12 mm	200–250 L of wine per stave	£7.50	£0.03+
Oak barrel chain	240 × 22 × 7 mm	1 per third or fourth fill barrel	£8.50	£0.037
Small tank staves	960 × 50 × 7 mm	50–60 L of wine per stave	£1.08	£0.02
Oak segments	50 × 50 × 7 mm 2.5 m² per bag	1500 L of wine per bag	£29.25 per bag	£0.02
Oak chips	7–10 mm	1.5–2 g/L	£1.95 per kg	£0.004
Oak granules	2–3 mm	1.5–2 g/L	£1.95 per kg	£0.004
Oak powder used with fermentation	Powder	1000 g/ tonne of grapes	£2.15 per kg	£0.003

Fig. 9.1 Inner staves inside an empty tank. To see a colour version of this figure, please see Plate 11 in the colour plate section that falls between pages 28 and 29

Of course, when wine is maturing in oak barrels, there is an interplay between the wine, the wood and the atmosphere: wine evaporates through the pores of the wood, and a controlled oxygenation takes place, together with the absorption of oak products. This does not happen when the wood is on the inside rather than the outside, such as the use of inner staves in a tank. Fig. 9.1 shows inner staves inside an empty tank. Micro-oxygenation, by which minute amounts of the gas are bubbled into wine in tank, is one way of attempting to emulate the breathing of wine in barrel. However, the use of oak chips, granules or powder is a poor substitute for ageing in oak barrels, often imparting incongruous sweet and sappy flavours.

9.2 Skills and diligence

It is easy to imagine that in the high-tech world of twenty-first century wine production, all involved are well-educated and highly skilled in matters relating to their particular field of work. This is usually the case at good quality producers, where the price commanded for the wine warrants, indeed necessitates, that every aspect is performed meticulously. At lower levels, lack of training, skills and diligence is often all too apparent. Vineyard work in particular is often carried out by lowly paid operatives who have had the bare minimum of training, including immigrant labour hired on a day-by-day basis. Without any sense of investment or ownership in the business and detached from the finished product workers underperform. A typical full-time vineyard worker in South Africa earns 350 rand per week (approximately £24.00 at the rate of exchange at the time of writing) per week, in a country where a litre of milk or a loaf of bread costs approximately 5 rand each. It is little wonder that some of the vineyards exhibit the lowest standard of viticultural work I have seen anywhere in the wine world an opinion endorsed by other wine writers. Pruning is a task that requires skill and a 'feel' for vine as a living being, but this task is often undertaken by pieceworkers who will work, at best, according to the simple formula instruction they have been given. Good fruit only comes from balanced vines, and sensitive pruning is one of the necessities to achieving balance.

9.3 Legal

It may seem strange that legal considerations may constrain quality, particularly within the European Union, until it is understood that many of the viticultural and oenological practices stipulated in the regulations were agreed after consulting, and perhaps under some pressure from, the large producers. It can be argued that the control of production costs, and thus control of production methods, was the prime driving force. Practices that are not in regulations as being permitted are forbidden. Small producers particularly often find that working within such a straightjacket of legislation restricts their ideas of improving quality, at least if they want to sell their wines as QWpsrs (quality wines produced in specific regions).

9.4 Environmental

In a wine world where priority is at last being given to measure to limit damage to, and if possible enhance, the environment, it may seem anathema that such considerations can have a negative impact on quality. Certainly, some environmental considerations have had a positive impact on quality. During the sixties, seventies and eighties, it seemed that one could not enter a vineyard without seeing routine preventative spraying or applications of yield 'enhancing' chemicals. Now a great many quality-conscious and environmentally friendly growers practise 'integrated pest management', and the last decade or so has seen a remarkable increase in the number of organic and biodynamic wine producers worldwide. Additionally, cover crops abound and the gentle hum of 'beneficial' insects is the main sound to be heard in the vineyard! Significantly, the standards of vineyard hygiene on the quality-conscious estate are higher than ever, the vines are better balanced and the quality of fruit has improved.

Today, the buzzwords are 'reducing the carbon footprint' and being a 'carbon neutral producer', and retailers and distributors are increasingly demanding that suppliers act to achieve this. It does, of course, provide very good public relations material. There is, however, a quality price to pay for demands to move to bottling or otherwise packaging wines in the country of sale rather than the region of production and in lighter-weight containers rather than glass bottles.

Historically, the shipping of wine in bulk was the norm, and bottling would take place by merchants in the country of sale. Wines have been shipped in barrels for centuries, but during the twentieth century other means of transport began to be utilised including, for cheaper wines, ships' tanks, SAFRAP (lined mild steel) containers and road tankers. Of course, producers had to prepare, adjust and protect wines for the journey, and the bottler would have to prepare the wines for bottling. There is no doubt that wines would often deteriorate during transport, and the integrity of the product was often compromised by the bottling merchant. Thus, in the second half of the twentieth century there was a growing move, especially for fine wines, to bottling at least in the region and at best at the property of production. Compulsory bottling at the property for *cru classé* Bordeaux wines was enacted for the 1972 vintage, although most chateaux had been bottling all or part of their production at the property for decades. Today, certain Appellation Contrôlée (and other national equivalent) wines must, by law, be bottled in the region of production.

A case of 12 × 75 cL bottles of still wine weighs approximately 15–16 kg, with 9 kg being the weight of the liquid and the rest being that of bottles, packaging, etc. There is also much 'wasted' space within the case, so shipping in bulk for bottling at destination, thus reducing both the weight and space required, makes prima facie environmental sense. As we have seen wines transported in bulk need to be stabilised for the journey – this may include an extra dose of sulfur dioxide – and after arrival will need further treatments to prepare for bottling. Moving wines, especially delicate whites, may result in a loss of freshness as is witnessed by examples of Chablis AC bottled not in the district but further south in the Côte d'Or.

The bottling of some inexpensive wines in lightweight plastic bottles that have something of the feel of their glass counterparts has recently been introduced. A glass bottle weights approximately 400 g, and the plastic bottles just 54 g. Plastic bottles may compromise for quality unlike glass which is a totally impermeable and inert material; plastic allows a small amount of oxygen migration. It also exhibits flavour absorption and wine acids can attack plastics. Sulfur dioxide levels have to be increased for wines bottled in plastic, with a resulting impact of loss of fruit. The life of the wine in plastic bottles is also reduced. The introduction of lighter-weight glass bottles is perhaps a more positive move, although there are some issues regarding their strength.

Production of Quality Wines

In this chapter we will consider some of the key factors in the production of quality wines. It must be stated that this book does not intend to be a manual of viticulture, vinification or oenology. Consequently, the list of topics covered is far from exhaustive.

10.1 Yield

The subject of yield, permitted and actual, and its impact on wine quality is a subject that arouses much discussion. The conventional view, particular prevalent in the Old World, is that restricting yield enhances the quality of the fruit produced. This concept is embodied in the laws of Appellation Contrôlée – the 'higher' the appellation, the more restricted the yield. There are several ways for growers to restrict yields including the density of planting, winter pruning, stressing the vines by deficit irrigation in countries where irrigation is practised and green harvesting. However, forcing low yields can be detrimental to quality, especially in vigorous sites.

Yields are usually expressed in hectolitres per hectare (or per acre), or particularly in Australasia, in tonnes per acre. Examples of basic permitted yields in the Bordeaux region are shown in Table 10.1.

There is no universal agreement as to what is the optimum yield for the production of high-quality wines, and most growers have to strike a balance between quality and economic production levels. Some growers will aim at the maximum possible permitted or achievable yield without concern as to the negative impact on quality. In Australia, unless top-quality wines are the objective, most

Table 10.1 Examples of basic permitted yields for Bordeaux ACs

Bordeaux AC – white (region)	65 hL/ha
Bordeaux AC – red (region)	55 hL/ha
Bordeaux Supérieur AC – red (region)	50 hL/ha
Haut Médoc AC (district)	48 hL/ha
Margaux AC (commune)	47 hL/ha
Sauternes AC (commune)	25 hL/ha

The basic individual yield stated (*rendement de basse*) may be varied each year (*rendement annuel*) by Institut National de l'Origine et de la Qualité. For example, in 2004 the annual permitted yield for Bordeaux AC (red) was 58 hL/ha.

growers will state that any yield below 4 tonnes per acre (approximately 10 tonnes/ha) is uneconomic. For red wines this equates to a yield of approximately 67–75 hL/ha, amounts in excess of the legally permitted yield of even the most basic (regional) Burgundy appellations.

The yield per vine is as important as the yield per hectare. A grower might claim a low yield, but if many of the vines are dead or dying, the yield per productive vine might be relatively high. The rules of Appellation Contrôlée provide that if the amount of dead vines comprises less than 20% of the total, the normal permitted yield per hectare applies; if the number is greater than 20%, then the permitted yield is correspondingly reduced.

10.2   Density of planting

The more vines per hectare, the more each vine competes for natural resources, including water. This is a limiting factor which, in the absence of rainfall, slightly stresses the vines. Vigour is restricted, and the energy of the plants goes into ripening a limited quantity of fruit. Having to compete for water encourages vines to send their roots deep, where they pick up minerals and trace elements.

Density of planting will depend on several factors, including historical, legal and topographical orientation as well as other aspects of the vineyard site, labour costs and the use of mechanisation. In France, in much of Burgundy and parts of Bordeaux, including vineyards of the Médoc, vines are planted at a density of 10 000 per hectare. The rows are set 1 m apart, and the vines are spaced at 1 m in the rows. Mechanical working takes place by using straddle tractors, which are designed to ride above vines with their wheels on either side of a row. This can give a problem of soil compaction – the wheels of the straddle tractor follow the same path with each pass. Regular tilling puts oxygen back into the soil, or alternatively the use of cover crops including cereals, legumes and grasses helps alleviate the problem.

At 10 000 vines per hectare, and based on a typical yield in the Haut-Médoc district of Bordeaux of 48 hL/ha, the average yield per vine is 0.48 L. In some New World countries the planting density might be as little as 2000 vines per hectare and a yield of 45 hL/ha would result in an average yield per vine of 2.25 L. There are areas, for example, Big Rivers in Australia, where a yield of 250 hL/ha was not uncommon (prior to 2006), and based on a density of 2000 vine per hectare, the production per vine might be 12.5 L.

10.3 Age of vines

It is generally accepted that young vines do not yield high-quality fruit, although as the vines are often balanced (see Section 10.4), occasionally the very first crops of some varieties such as Sauvignon Blanc or even Pinot Noir can produce some classic varietal flavours. It is also generally accepted that old vines produce the finest wine grapes, albeit in a lesser quantity. After an age of approximately 20 years, the vigour of the vines is reduced and balance is restored. However, there is no consensus at what age vines become mature and at what age old. It is not uncommon to see the words 'vieilles vignes' on labels of French wine, but the term has no legal definition. Many top producers, including Bordeaux crus classes, exclude the crop taken from young vines from their grand vin. Some define 'young' as less than 7 years old others as less than ten. On the steep slopes of the Dr Loosen Estate in the heart of Germany's Mosel region, Ernie Loosen replaces dead or dying vines on a one-by-one basis and thus maintains a particularly high average vineyard age, a factor that he believes is a major contributor to the acclaimed, exceptional quality wines.

10.4   Winter pruning

For most growers this is the main way of restricting yield. Pruning takes place when the vines are dormant. The more nodes that are left on the vine, the greater the potential yield. The range is between 6 and 40 nodes. Some growers have a fairly rigid formula, e.g. a vine will be pruned to 10 nodes. Others prefer to consider the vines as individuals and decide on an appropriate charge according to the age, health and vigour of each vine. The aim should be to achieve vine balance, i.e. the balance between fruit and vegetation. A simple way to measure this is to weigh the prunings and compare with the weight of fruit yielded by the vine in the previous harvest. The weight of fruit should be between five and ten times the weight of prunings – more than this and the vine is overcropping (and should be pruned harder), less than this it is too vigorous and more nodes should be retained. Careful pruning and training (including subsequent summer pruning, removal of laterals and leaf plucking) also helps maintain a healthy canopy, reducing the likelihood of cryptogrammic diseases such as **powdery mildew**, also known as **oidium** (*Erysiphe necator*), and **downy Mildew**, also known as **peronospera** (*Plasmopara viticola*).

10.5   Stressing the vines

Vines, like people, work best when slightly stressed. This is a conventional view that is now challenged. Obviously a grower has no control over rainfall, but in countries and regions where irrigation is permitted, the amount of water given to vines has a major impact on quality. Growers may practise deficit irrigation. Neutron probes in the soil indicate when water is required. Stressing the vines causes the roots to synthesise abscisic acid, sending this to the leaves and deceiving them into reacting as though there are drought conditions. Shoot growth stops and all energy goes into ripening the fruit. Moderate water deficit can double or treble the concentration of the precursors of the varietal thiols that will be released during fermentation. Partial root drying is a very recent technique. The roots on one side of the vine are given insufficient water, while the other side is irrigated normally. Before any damage occurs to the roots, the irrigation pattern is reversed. However, as with many viticultural matters, the concept of vine stress and deficit irrigation as a vehicle to improve quality is far from universally accepted. For example, Neal Ibotson of the multi-award winning Saint Clair Estate in the Marlborough region of New Zealand states that stressed vines give poor quality, a view endorsed (particularly in the case of Sauvignon Blanc) by Kevin Judd, chief winemaker at Cloudy Bay vineyards.

Quality is affected not just by whether the vines are stressed, but when they are stressed. As we have seen, it is beneficial for Sauvignon not to be stressed, especially if pyrazine aromas are desired in the wine. Unstressed Cabernet Sauvignon also produces very pyrazine-dominated wines. Whilst with Sauvignon Blanc this is considered desirable, it is not with Cabernet Sauvignon, particularly as the tannins are also adversely affected by excess water.

10.6   Green harvesting

This involves removing fruit from the vine before ripening. Having ensured that the vine is bearing at least the desired amount of fruit (having endured any early season climatic, disease and pest risks that might have limited the yield), a green harvest of fruit in excess of the desired yield will allow the energy of the vine to go into fully ripening a lesser quantity of high-quality fruit. Flavour as well as sugar concentration is improved. However, many viticulturalists challenge the concept, believing that if green harvesting is necessary, the vines are out of balance.

10.7   Harvesting

The timing and method of harvesting have a major impact on quality. A grower or producer's decision as to when to harvest will be decided upon with respect to many factors, including seasonal parameters, the weather forecast, the risk of botrytis, the balance of sugar and acidity in the grapes, the maturity of skin and seed tannins and the formation of aromas and aroma precursors. Of course, as the harvest cannot be organised at a moment's notice, careful analysis and sophisticated prediction tools are employed.

Historically, especially in some regions of the Old World, many growers would pick early: though picked too early for the grapes often lacked phenolic ripeness or even the desired levels of sugars. Growers also lived in fear of autumn rains spoiling the harvest with bunch rot and dilution and so undertook early 'insurance' picking. In order to ensure that the grapes have an acceptable, minimum level of ripeness, the rules of Appellation Controlée include, in most regions, a *'ban de vendage'*, a date variable each year before which the harvest may not take place. Nowadays, however, growers more often wait for phenolic ripeness which, in warm and hot climates, may not come until after the desired levels of sugar and acidity are present. This can result in over-alcoholic and over-loud wines. Blair Walter of Felton Road Winery in New Zealand's Central Otago region states that he is making a conscious effort to pick earlier and is not afraid of a green edge. Jeff Synnott of nearby Amisfield Wine Company agrees: 'Our notion of ripeness and quality has somehow become confused'.

10.7.1 Mechanical harvesting

Some viticulturalists claim that in order to pick the right fruit at the right time machine harvesting is essential. Mechanical harvesters cost between £50 000 and £125 000, the price being partially dependent on whether they are self-propelled or towed. A machine can pick in 1 day the quantity of grapes that would require 40–80 manual pickers. The technology of mechanical harvesters has improved considerably during the last 20 years. The harvesting system generally now used is that of bow-rod shakers. The fibreglass rods move the fruiting zone of the vines to the right, and then rapidly to the left, and using the principle of acceleration and sudden deceleration of a mass the grapes are easily snapped off the stalks, which are left as skeletons on the vines. Lightweight berries, including shot berries, raisins and those that have been bird eaten do not have the mass, and so are left behind. The fruit is initially collected on conveyers comprising interconnected shallow baskets which are moving in a rearwards direction at the same speed as the harvester is moving forwards. MOGS (matters other than grapes) are removed by two methods: spinning blades just above the trays will remove large MOGS, and vacuum systems remove light matters such as leaves.

Advocates of mechanical harvesting cite the advantages as follows:

(1)   Machines work fast and the grower can utilise them at optimum ripeness.

(2)   The fruit picked is cleaner than may be the case with unskilled labour.

(3)   Machines pick at a controlled rate so there should be no delay in processing, which gives an advantage over manual picking which can result in queuing for the crusher and consequent fruit deterioration.

(4)   Machines can work at night and deliver desirable cool fruit to the winery.

(5)   Cost of picking is generally less than half that of using manual labour although the cost of the machine has to be amortised.

10.7.2   Hand picking

In spite of considerable improvements during recent years, machines do damage grapes to a small extent. Trained pickers can select fruit according to requirements and pick into small bins, sometimes just 12 kg capacity, which will also minimise fruit damage. The use of mechanical harvesters is, of course, not possible if the vineyard has steep slopes (above 20–30% depending on row orientation) or if the trellis system is not suitable, in which case hand picking is the only option. Hand picking is also the only option if the winemaker wishes to vinify separately the product of small parcels of individual terroirs. Hand picking is essential if whole clusters are needed, as for certain white wines and Beaujolais. It is also a legal requirement for Champagne and, of course, Sauternes where part bunches or even individual berries of grape affected by *Botrytis cinerea* in the form of 'noble rot' are desirable.

10.8 Delivery and processing of fruit

Ideally, the fruit should be transported to the winery in small containers, so it does not crush itself under its own weight. Grapes should be processed as soon as they are harvested – some top producers aim to have fruit in the crusher within 1 hour of picking. The longer the grapes are left waiting before they are processed, the more risk there is of bacterial spoilage or oxidation. This is particularly important in a hot climate, when the process of deterioration is much faster than in a cool climate. Insects, including *Drosophilidae* (vinegar flies), are immediately attracted to damaged fruit.

10.9 Selection and sorting

The purpose of sorting is to remove unripe fruit and altered fruit, that is fruit affected by rot. Sorting can take place at a number of points:

(1) Selection in the vineyard – pickers may be instructed which fruit to harvest and which to leave on the vines.

(2) A sorting table on a trailer will be situated at the ends of the rows of vines, and the fruit will be sorted manually by people on each side of a moving belt.

(3) Upon arrival at the winery clusters or, in the case of machine harvesting, berries, sorted manually by people on each side of a moving belt.

(4) After de-stemming, in the case of hand-picked grapes.

There are now several designs of mechanical sorters working on the principle of the differing masses of healthy and damaged grapes, and these can be particularly valuable where the cost or skills of labour are a major issue.

10.10   Use of pumps/gravity

Pumps are, of course, used extensively in wineries for tank filling, pump-overs (if practised), racking, bottling, etc. They are a physically aggressive way of moving liquids and their use for moving must can be particularly detrimental, breaking seeds and causing an increase in dissolved oxygen. Wineries should be designed or adapted to enable more gentle movement by gravity as far as possible. For instance, fruit from the crusher can be discharged into a small tank which is then lifted by hoist to gravity feed the fermentation vats, or in the case of white wine, the press. In the production of red wines, if the fermentation vats are high enough above the ground, at pressing time a horizontal press or the cage of a vertical press may be moved to the discharge hatch of each vat and filled with the minimum of movement of the skins. Figures 10.1 and 10.2 show the excellent gravity feed winery at Rustenberg Estate, Stellenbosch, South Africa.

10.11   Control of fermentations

It is important that temperature is controlled throughout the fermentation period. Although temperature control systems for large fermentation vats are now commonplace, challenges are still posed in adjusting the ferment to the desired temperature (normally cooling but occasionally warming) for particular stages of fermentation. Temperature is not uniform throughout the vat – heat rises and regular equalising of the ferment is desirable. Barrel ferments, although having a greater percentage surface area for heat discharge, can present problems too, and an ambient cooling system is highly desirable. White wines generally require fermentation under cool conditions (12–16°C) especially if an aromatic style is required, or slightly warmer (17–20°C) for full-bodied styles.

Fig. 10.1 Gravity feed winery at Rustenberg Estate, Stellenbosch. To see a colour version of this figure, please see Plate 12 in the colour plate section that falls between pages 28 and 29

Winemakers fear a fermentation 'sticking' – stopping before all the sugar has fermented out. This can happen because of several reasons. The temperature may have exceeded that at which yeasts will work (approximately 35–38°C), or the juice may be deficient in oxygen or nutrients. However, for good colour and flavour extraction

Fig. 10.2 Gravity feed winery, showing press, at Rustenberg Estate, Stellenbosch. To see a colour version of this figure, please see Plate 13 in the colour plate section that falls between pages 28 and 29

for reds, the winemaker may wish to ferment at 30–32°C, particularly during the early and middle fermentation periods. The tolerance range below the sticking point is not very wide and examples of cooked flavours from overheated ferments are all too common. Stuck ferments are much less of a problem than in the past when hot autumns posed particular challenges. There are many recorded instances of stuck fermentations at *cru classé* properties in Bordeaux before the advent of efficient temperature control systems, spearheaded in the 1960s at Château Haut Brion and Château Latour, but not commonplace elsewhere until the 1970s. There are still properties lacking temperature control equipment in vats made of cement.

10.12   Use of gases

At many stages during the winemaking process, inert gasses may be used to maintain freshness and prevent oxidation and other spoilage. The gases commonly used are nitrogen, carbon dioxide and, increasingly, argon. Carbon dioxide is highly soluble in wine, and its use can give the finished wine a 'prickle' on the tongue. This is particularly undesirable in red winemaking. Nitrogen, however, having a lower molecular weight does diffuse quickly. Accordingly, mixtures of two of the gasses have advantages.

Horizontal tank presses and fermentation vats may be sparged before filling. Partially filled vats may be blanketed by gas and at racking and bottling times empty vessels may also be flushed with gas. Considerable amounts are required to flush all the air from the vats. The average quantity of gas used in producing a (9 L) case of Australian wine is 100 L. For high-quality wine this rises to 600 L.

10.13   Barrels

For those wines that benefit from oak ageing, there is no doubt that barrels are infinitely superior to the oak treatments discussed in Chapter 9. The impact of the barrels on the style and quality of

Table 10.2 Cost of oak barrels

Size	Type	Type of oak	Price	Cost/litre
225 L	Bordeaux export	French	£440	£1.95
225 L	Bordeaux export	American	£265	£1.18
225 L	Chateau	French	£535	£2.38
300 L	Transport	French	£504	£1.68
500 L	Transport	French	£738	£1.48
300 L	Transport	70% French, 30% Euro	£434	£1.45

wine will depend on a number of factors including size of the barrel; type and origin of oak (or other wood); manufacturing techniques including toasting; amount of time spent in barrel; and where the barrels are stored. There is also a marked difference between the maturations of the so-called château barrels, using fine-grained oak that has a thickness of 22 mm, against the standard 'export' barrel with a thickness of 27 mm. Most quality-conscious producers prefer to source their barrels from several coopers, as each has their 'house style' which will impart subtle differences to the maturing wine and give a greater range of components to influence the quality of the final wine.

The approximate costs of new 225 L barriques in 2007 are detailed in Table 10.2.

Wine put into barrel for maturation generally requires regular topping up, although some oenologists question the effectiveness of this exercise. The barrels will require regular racking – perhaps six times for wines that spend 18–21 months in barrel.

10.14 Selection from vats or barrels

The production of top-quality wine involves making selections throughout the vinification process – from arrival of fruit at the winery to the components of the finished wine. The exclusion of vats or barrels containing wine from younger vines, or wine from poorer parts of the vineyard, or that which simply has not turned out to be of a very high standard, is most desirable. The excluded wines can be sold as a 'second wine'. Alternatively, they may be or moved down a level down in a multi-tiered range of wines or, if substantially inferior, sold to a bulk producer as a blending component.

10.15   Storage

Whatever be the quality of wine immediately after bottling, how the bottles are subsequently stored will significantly impact on the quality at the point of opening. Some wines have a legal minimum period of ageing in bottle before being released for sale; e.g Rioja DOCa Gran Reserva (red) must be bottle-aged for at least 2 years (with at least 5 years total ageing including at least 2 years in cask). However, many producers wish to give their wines at least a

---

The conditions needed for bottled wine storage

Darkness	Apart from the fact that light is usually also a source of heat, exposed bottles can suffer from 'light-strike'. White wines will darken and red wines become more brown.
Laying down	Bottles should be stored on their sides in order to keep the corks moist and thus expanded in the neck of the bottle. Wines with plastic or screw-cap closures should be kept upright.
Constant temperature	Not always easy to achieve. A constant temperature promotes a controlled and balanced maturation. Corks that expand and contract with temperature changes are at risk of failure.
Temperature of 13°C (55°F)	This is often referred to as cellar temperature. A couple of degrees discrepancy either way will not pose a major problem. Wines mature quicker and less evenly if the temperature is much above this. **Constancy is more important than the actual temperature.**
Away from strong smells	These can be absorbed through the closure.

Fig. 10.3   Conditions needed for bottled wine storage

short period of time in bottle before release from the cellars, even if not legally required to do so.

Storage conditions for wines laid down for bottle maturation need to be particularly suitable, but even wines designed for immediate drinking can be irreparably damaged if exposed to excesses of light, heat or cold. Details of suitable storage conditions are shown in Fig. 10.3.

CHAPTER 10

Selection by Buyers

In this chapter I will look at some of the considerations buyers may have to take into account when selecting wines for resale.

Historically, wine was the domain of wine merchants who would supply trade customers and/or public from stocks sourced from long-established contacts. France and Germany dominated their lists and lists that were for the more expensive wines would be hidden from the customer's sight. Most customers, not just the novice, would likely have felt intimidated when faced with the gentleman in the pin-striped suit who clearly had vastly superior knowledge. Today, it is very different. Whilst there are some excellent regional merchants, their strength lies in wines that cost £6 or more, and it is the supermarkets who dominate the volume market. Supermarkets de-snobbed wine, made buying easy and women friendly, and rocketed sales. However, producer/suppliers now lament that although the overall UK market achieved decades of continuous and consistent growth (at least until 2006 when it stalled), the number of important trade buyers of everyday wines diminished. It is now a number that can almost be counted on the fingers of one hand, due to mergers, acquisitions and the incessant rise in supermarket power and market share.

11.1 Supermarket dominance

To the wine lover, the image is an endearing one. The supermarket buyer is travelling from one small wine estate to the next, sidestepping chickens in the yard and the obligatory vineyard dog, to taste the wines as the anxious host looks on. Suddenly comes the wide beam on the buyer's face, the 'eureka' moment, and he proudly informs the beaming owner – 'this is the one – I'll take all you have'.

In the real world things are somewhat different. There is a joke amongst wine producers that a supermarket buyer is assessing one of the wines in particular detail. The nose goes in and out of the glass; the wine is swirled, tasted again, followed by that long reflection. There is some scratching of the head. Finally comes the pronouncement: 'sorry – there's just not enough margin!' Whether or not a wine is listed will depend not just on the actual product and sometimes an audit of the producer's facilities, but also on listing allowances, pre-investment, promotional budgets, exposure funds, market support, supplements for gondola ends, retrospective discounts, etc.

In the UK, according to *Nielsen*, supermarkets account for 68% of the total sales by value and over 70% by volume. However, the figures are subject to challenge as *Nielsen* only records electronic point of sale data. This perhaps gives a very distorted picture of the UK market as they include almost all supermarket sales, but very little of the sales of specialist wine merchants, leading one to believe that:

(1)   there is little market for fine wine;
(2)   everyone gets most of their wine from supermarkets;
(3)   the average price per bottle is very low;
(4)   big-volume brands dominate the market.

'The public wants what the public gets' is a line from a 1970s popular anthem by the band Jam and this has never been more true, particularly insofar as the unimaginative range of many supermarkets is concerned. Twenty years ago the real diversity of wines offered was much greater than today, even though now the supermarkets boast of having more labels. Let us take south-west France as an example. At that time the UK shopper could easily find some eclectic and distinctive wines. For instance, Jurançon AC, sweet spicy and tropical, made from the Petit and Gros Manseng grape varieties; white Gaillac AC, somewhat nervy, made from a blend of mostly local varieties including Len de l'El, Mauzac and Ondenc; Madiran AC with the wonderful leathery anis flavours of the predominant Tannat; Irouléguy AC, deep, dark and tannic and again made mostly from Tannat; and many more. Today, their place has been taken on the shelves by ubiquitous south-eastern Australian Chardonnay and Cabernet Shiraz, perhaps bearing many different labels including the supermarkets' own, but in reality the bottles contain similar wine as they are produced by one or other of the 'super' wine factories. The reds are fruity and soft. The whites are fruity and soft. Classic regions too have lost listings, for example, the number of wines from Bordeaux listed by supermarkets is very small compared with the production and worldwide importance of the region.

Of course, supermarkets do list many good quality wines that show regional identity, but buyers often have safety in mind. They avoid wines that have such misunderstandable characteristics that might result in or returns, as witnessed by a buyer's aide-memoir on the wall of a supermarket's tasting room: 'reject wines with sediments' and 'reject garnet or tawny reds', even though such wines could be classic examples. It should be noted that some supermarkets do have a particularly fine, if small, 'inner cellar selection' of top-quality wines available in flagship stores.

11.2   Price point/margin

There are many crucial price points for wines, and producers and retailers break through them at their peril. The selling price point and profit margin thereon is therefore a major consideration when selecting potential listings. It is a sad, even depressing, fact that at the time of writing, and after the swingeing excise increases in the 2008 budget had been applied, according to *Nielsen* (and the reader should note the caveat above) the average selling price for a 75 cL bottle of wine in a UK supermarket was £3.90. Multiple specialist merchants sold at an average price of £4.94 per bottle, but taking the off-sales market as a whole, that is sales other than in pubs, wine bars, clubs and restaurants, the average price was still only £4.08 (again revealing the dominance of supermarkets). It is interesting to note that the country with the highest average price (£6.44) per bottle is New Zealand, relatively lowly placed at number 9 in the UK sales league table, but with high consumer perception of quality, particularly of wines made from Sauvignon Blanc and Pinot Noir. With regular rises in excise duty, which retailers expect suppliers to fund, the pressure on producer margins is greater than ever. Retailer margins are healthy: it is not possible to quote an average figure but many retailers achieve 25–35% profit on return. The on-trade, of course, looks for much greater margins, and the published accounts of one large group of public houses and gastro-pubs (which offers a wide selection of wines by the glass) reveal that they are working on 55% profit on return.

11.3   Selecting for market and customer base

Independent wine merchants have often been heard to say when choosing an outstanding but obscure wine for their list: 'if I can't sell it I'll be happy to drink it'. Partly due to the internet giving the independents a possible national rather than restricted local market, the number of independents is again increasing after years of decline. However, for many independent wine merchants knowing their customers, understanding their aspirations and selecting wines that fit with these are the keys to success. Multiple specialists too have researched their customer profiles and even brand outlets to suit. Wines are chosen to reflect the customers' perceived needs and spending patterns. Restaurant lists, often written by their wholesale supplier, need to have regard for the client base, the profile and, of course, the menu.

11.4   Styles and individuality

Any wine list should exhibit a broad range of styles and prices. At low price points for 'entry level wines' to use the marketers parlance, the concept of individuality is non-existent, but the economies of scale make possible clean and fruity wines for those to whom wine is little more than a commodity. For wines made from a single variety the character should be apparent, and dual-variety blends should express both components as well as constitute a harmonious whole. It is true to say that overproduction of grapes, as is presently the case in many countries of both the New World and Europe, has resulted in a plentiful supply of sound fruit for such wines. Even if the label states a district of origin, this does not guarantee that the wine will show the identity of the area. It is unrealistic to expect that, for example, a Minervois AC retailing at little more than £3 per bottle will be anything other than a simple drink, but with a hint of the character of the Languedoc-Roussillon region. However, even at relatively low price points, fraudulent labelling does still exist. Historically, it was the illustrious appellations whose wines commanded high prices that were often compromised. Today, it is fashionable varieties such as Pinot Grigio from popular regions such as Veneto that are likely to be the subject of what I can only describe as counterfeiting. Far more Italian 'Pinot Grigio' is sold than the country produces!

As we start to progress up the price scale, there is a broadening of available styles and the possibility of regional, district and producer identity. The price at which real regional characteristics become apparent depends very much on the region in question. For example, it is possible to find Vouvray AC that retails at around £6 that shows classic Chenin Blanc peach character, lively and racy with the classic mineral tones of this part of Touraine. A true-to-type Pouilly Fumé AC, produced further up the Loire in the Central Vineyards area and should be flinty, gently smoky, crisp and elegant with restrained Sauvignon Blanc nettle and gooseberry character, will have a starting price of £10 or more. From around this price point wines should express true individuality, they should excite and stimulate, and they should say who they are, why they are and precisely where they are from.

11.5 Continuity

Many consumers today expect products to be continuously available, and indeed the very concept of seasonality, the eager anticipation of the new season's lamb, cherries or peas hardly exists. This applies to wines too, and the excitement of drinking Beaujolais Nouveau in November from grapes that had still been on the vine in September is long gone. For the retailer or restaurant too a lack of continuity of a product can present a major headache, with increased costs in changing lists, etc.

Historically, wholesalers would hold considerable stocks of both young and mature wines and could, to a large degree, balance supply to their customers. Today, producers increasingly supply larger trade customers directly or use the services of an agent or sales office who holds no stocks. But modern customers require continuity of supply and stability of price. This may be planned, and financed, within the available stocks of a vintage, but it is unrealistic to expect such guaranteed continuity from year to year. Wine is an agricultural product and as such is at the mercy of the weather, diseases and other natural forces that can result in considerable annual variation in both the quantity as well as quality. This is particularly true in the Old World, but New World countries are not immune. For example, the 2007 total wine grape harvest in Australia, following a growing season affected by continuous drought, frosts and bushfires resulting in smoke taint, was 1.39 million tonnes. This figure was down from 1.9 million tonnes in 2006. This equates to a drop in the number of litres of wine produced from 1410 million litres in 2006 to 955 million litres in 2007, a reduction of over 30%. Shiraz, the most widely planted red variety, and Cabernet Sauvignon each suffered a 36% drop in crop. As there was considerable surplus wine being kept in tank from previous vintages, the impact on the amount of product available for release to the market (especially as far as 'commodity' wines were concerned) was limited, but the dangers as far as continuity of supply to a market that has become accustomed to key price points are apparent.

11.6   The place of individual wines in the range

When selecting wines for inclusion in a range, the buyer needs to have in mind the other wines already listed and, unless existing wines are to be directly replaced, how the new selections will sit against them. It is unlikely that the consumer will buy more wine simply as a result of extra listings, and it is probable that each wine added will take the sale from one already stocked. At the higher price points each wine will need to have not only quality but also individuality and to earn its place on a list by saying something that its peers do not.

11.7   Exclusivity

It is easy for any retailer or restaurant to build a list around large and well-known brands. Such lists might be comforting, but lack identity and interest, and will lack any competitive advantage, other than possibly on price. Selecting individualistic wines that competitors do not have creates interest, and can give competitive edge and healthy margins, for the customer has no direct price comparisons. This can be very important for restaurants whose customers resent paying three times the price for a wine readily available on the supermarket shelves. Sourcing and retaining lines that are exclusive to the retailer (always staying within anti-competition legislation rules) can prove challenging but rewarding. Producers will often prepare an alternative label for those buying sufficient quantities or the retailers may wish to have ownership of their own label. Buyers' Own Brands (BOBs) are now a very important means of guaranteeing exclusivity and, of course, the wine may be prepared to the customer's specification. It is worth noting that producers too may wish to offer an alternative label as a means of disposing of excess stocks at a price lower that their usual label without devaluing its status.

11.8   Specification

Producing wines to buyers' specifications may seem to be a very modern phenomenon but this has, in a much less technical way, been taking place for centuries. Put simply, a producer would want to make the product as appealing as possible in the market in which it was to be sold. For example, the wines of Bordeaux have long been exported to Britain, a trade well established in the 300 years (1152–1453) that Bordeaux was English. However, customers often found the wines too light for their palate (the name claret used for the red wines of the region comes from a French word *clairet* meaning pale red). Accordingly, the tradition grew of blending in some deeper and fuller-bodied wine from the warmer Rhône Valley, a practice known as *hermitagé*, although it is unlikely that the additive came from the great vineyard of Hermitage. Also, before the wines were shipped in cask, they would have a little spirit from the neighbouring region of Cognac added to give them an alcoholic boost and to help stabilise them for their sea journey. Thus, the style of wine was adjusted for the market. Burgundy too was often adjusted for the *goût anglais*. The railway station at Beaune became an important link in the production chain, for here were received bulk quantities of cheap full-bodied wines from warmer regions further south, particularly the Rhône Valley, to be blended in with the often thin wines from vines that were over-yielded. Thus, the wines were adjusted to the specification of the buyer (full body) and the seller (reduced production cost).

Today, the supermarket and multiple specialists, especially when considering a buyer's own brand, may have a much more detailed specification as to style and aspects of technical composition, particularly regarding the levels of alcohol, acidity and residual sugar. With modern production techniques in the large wineries, fulfilling the requirements presents few challenges. For example, the demand for deep-coloured, red wines with soft tannins at low price points has coincided with technology that make these achievable including flash détente and thermo détente. Containers and closures are specified too – in the UK several supermarkets specify that their producer/suppliers use screw-caps or other synthetic closures.

11.9 Technical analysis

Until recently there were many thousands of small producers throughout the world who never submitted their wines to more than a basic analysis. The wines from some of these were truly excellent, and from others distinctly mediocre. Perhaps the only measurement equipments used during the winemaking process were weighing scales, capacity measures, a thermometer and some hydrometers. By differential calculation of the specific gravities, that is of the start must weight and the wine after fermentation, the percentage alcohol of the finished wine was determined to the accuracy needed for labelling.

All wines now require at least a basic laboratory analysis to ensure that they comply with the legislation in the country of production and sale, including specific requirements for QWpsrs (quality wines produced in specific regions) if applicable. Interestingly, European Union (EU) regulations require that QWpsr are subject to tests for 'wine *behaviour*' (author's italics). At the time of writing the only 'technical' information required on the label of a wine sold in the UK is the content, the percentage alcohol to an accuracy of 0.5% and, since November 2005, the allergen information that the wine contains sulfur dioxide (sulfites). EC Directive 2007/68 adds egg and milk products (commonly used as fining agents) to the allergen information to be included on labels with effect from May 2009. However, the industry may argue for milk products to be excluded. Generally, ingredient labelling is not required. From a marketing point of view this may be a good thing for much as wine is perceived as a natural product, EU regulations detail 50 permitted additives (the permitted list in USA totals 65). There are, of course, legal limits on the amount of the many chemicals that may be present in or added to a wine specified in EU regulations (Note: further restriction exists in some member states). Third countries have their own regulations and without detailed analysis producers are exposed to the risk of non-compliance. Some of the chemicals present may have been used in the winemaking process, some derive from the grapes or are manufactured by the wine itself, and others are contaminants. For example, Ochratoxin A (OTA) is produced by some species of fungi that infect many crops including cereals, coffee and grapes. The toxin can cause liver and kidney damage, and is a possible carcinogen. EU regulation 123/2005 (which amends 466/2001) specifies a legal OTA limit in wine of 2 μg/L. EU regulation 401/2006 lays down sampling procedures for mycotoxins in foodstuffs which, of course, include wine.

Table 11.1 Parameters used in the analysis of wine

2,3,4,6-Tetrachloroanisole – ng/L
2,4,6-Tribromoanisole – ng/L
2,4,6-Trichloroanisole – ng/L
4-Thylphenol – µg/L
Acetaldehyde – mg/L
Alcohol – % vol. at 20°C[a]
Arsenic – mg/L[a]
Ascorbic acid – mg/L[a]
Bacteria – CFU (colony-forming units)
Bitterness – bittering units
Brettanomyces spp.
Calcium – mg/L[a]
Calorific value – kJ/100 mL (kCal/100 mL)
Carbon dioxide – g/L
Cold stability[a]
Contents – mL
Copper – mg/L
Copper – mg/L[a]
Density – g/L
Discolouration
Dissolved oxygen – mg/L
Ethyl carbamate
Filterability
Free sulfur dioxide – mg/L[a]
Fructose –g/L
Glucose – g/L
Glycerol – g/L
Haze – formazine turbidity unit
Hydrogen sulfide – µg/L
Iron – mg/L[a]
Lactic acid – g/L
Lead – mg/L[a]
Magnesium – mg/L
Malic acid – g/L
Ochratoxin A – µg/L
Optical density – 420 and 520 nm
Pentachloroanisole – µg/L
pH[a]
Potassium – mg/L[a]
Protein stability[a]
Reducing sugar – g/L
Silver – mg/L
Sodium – mg/L[a]
Sorbic acid – mg/L[a]
Specific gravity[a]
Sugar-free dry extract – g/L[a]
Total (titratable) acid – g/L[a]

Table 11.1 Parameters used in the analysis of wine (*Continued*)

Total dry extract – g/L
Total phenolics – mg/L
Total potential alc. – % vol.
Total residual sugar – g/L[a]
Total sulfur dioxide – mg/L[a]
Volatile acidity – g/L (acetic[a] acid)
Yeast – CFU (colony-forming units)

[a] A certificate showing analysis results for items marked are commonly required by United Kingdom supermarkets.

When comparing the results of analysis, it is important to ensure that identical expressions of the measurement are used. For example, total acidity is usually expressed as g/L tartaric acid, but in France and some other countries it is often expressed as g/L sulfuric acid (to convert a tartaric acid figure to sulfuric it is necessary to divide by 1.531). Table 11.1 lists analytical parameters.

Appendix

WSET® Systematic Approach to Wine Tasting (Diploma)

CHECKLIST: EXAMPLES OF TASTING TERMS

APPEARANCE

Clarity		bright – clear – dull – hazy – *(faulty?)*
Intensity		water-white – pale – medium – deep – opaque
Colour	white	lemon-green – lemon – gold – amber – brown
	rosé	pink – salmon – orange – onion-skin
	red	purple – ruby – garnet – tawny – brown
	(rim and core)	
Other observations		legs/tears, deposit, petillance, tints/highlights

NOSE

Condition	clean – unclean *(fault: oxidised – out of condition – cork taint – other)*
Intensity	light – medium(-) – medium – medium(+) – pronounced
Development	youthful – developing – fully developed – tired/past its best deliberate oxidation?
Aroma characteristics	fruit – floral – spice – vegetal – other

WINE QUALITY

PALATE

Sweetness		dry – off-dry – medium-dry – medium – medium-sweet – sweet – luscious
Acidity		low – medium(-) – medium – medium(+) – high
Tannin	level:	low – medium(-) – medium – medium(+) – high
	nature:	eg ripe/soft vs unripe/green/stalky, coarse vs fine-grained
Alcohol level		low – medium (-) – medium – medium(+) – high fortified (low/medium/high level?)
Body		light – medium(-) – medium – medium(+) – full
Flavour intensity		light – medium(-) – medium – medium(+) – pronounced
Flavour characteristics		fruit – floral – spice – vegetal – other
Other observations		eg texture, balance
Length		short – medium(-) – medium – medium(+) – long

CONCLUSIONS

Quality	faulty – poor – acceptable – good – outstanding
Reasons for quality	eg balance, concentration, complexity, length
Origins/variety /theme	eg - location (country, region) eg - grape variety/varieties eg - production methods/climatic influences etc
Price	(approximate retail price):
Age (in years)	(age in years):
Readiness for drinking/potential for ageing	needs time *(how long?)*– ready to drink, but can age *(how long?)*– at peak/drink soon – declining – tired/past its best

AROMA AND FLAVOUR CHARACTERISTICS

FRUIT

Citrus	grapefruit, lemon, lime
Green Fruit	apple (green/ripe?) gooseberry, pear
Stone Fruit	apricot, peach
Red Fruit	raspberry, red cherry, plum, redcurrant, strawberry
Black Fruit	blackberry, black cherry, blackcurrant
Tropical Fruit	banana, kiwi, lychee, mango, melon, passion fruit, pineapple
Dried Fruit	fig, prune, raisin, sultana

FLORAL

Blossom	elderflower, orange
Flowers	perfume, rose, violet

SPICE

Sweet	cinnamon, cloves, ginger, nutmeg, vanilla
Pungent	black/white pepper, liquorice, juniper

VEGETAL

Fresh	asparagus, green bell pepper, mushroom
Cooked	cabbage, tinned vegetables (asparagus, artichoke, pea etc), black olive
Herbaceous	eucalyptus, grass, hay, mint, blackcurrant leaf, wet leaves
Kernel	almond, coconut, hazelnut, walnut, chocolate, coffee
Oak	cedar, medicinal, resinous, smoke, vanilla, tobacco

WINE QUALITY

OTHER	
Animal	leather, wet wool, meaty
Autolytic	yeast, biscuit, bread, toast
Dairy	butter, cheese, cream, yoghurt
Mineral	earth, petrol, rubber, tar, stony/steely
Ripeness	caramel, candy, honey, jam, marmalade, treacle, cooked, baked, stewed

Glossary

Acetaldehyde. Product of the oxidation of ethyl alcohol; at low concentrations enhancing aroma, but at high concentrations giving a 'sherry-like smell'.

Anthocyanins. A group of polyphenols, found in the skins of red grapes, (and many other fruits and flowers) that is mainly responsible for the colour of red wines.

Autolytic. The character resulting from an interaction between wine and solid yeast matter, resulting in a distinctive bread- or biscuit-like flavour. Autolytic character is encouraged by ageing a wine on the lees, as this takes place in the bottle during producer maturation of Champagnes or in vat or barrel for some Muscadets and certain other whites.

Bentonite. A type of clay that can be used for fining (q.v.). Bentonite has many other uses including as a lubricant for drilling bits in the oil industry, tanking of ponds and as an enema.

Beerenauslesen. A category of German wine made from a selection of individual overripe or botrytised berries. The grapes are so suger-rich it is impossible to ferment to dryness, resulting in wines that are very sweet. It is only possible to make the wines in very good years. Trockenbeerenauslesen are made from grapes that have become dehydrated, like raisins.

Botrytis cinerea. Fungus that can attack grapes in the form of destructive grey rot, or in certain climatic and weather conditions as 'noble rot', desirable for the production of sweet wines such as Sauternes and Beerenauslesen (q.v.).

Chaptalisation. The addition of sucrose or concentrated grape juice to the must or the juice in the early stages of fermentation to increase alcohol level. In cooler climates grapes often do not contain enough sugars to produce a balanced wine. The process is named after Dr Jean Antoine Chaptal who was minister of the interior at the time of Napoleon.

Charmat. Sparkling wine production method, used for inexpensive wines, in which the second fermentation takes place in tank. Named after its inventor.

Colony-forming unit (CFU). A measurement, normally per millilitre (in the case of liquid food products) of the number of viable cells of micro-organisms (bacteria, yeast and fungi). NB: A single cell can reproduce by means of binary fission.

Cru. Lit. growth. A French term for certain ranked vineyards or communes.

Cru classé. An officially classified Bordeaux wine-producing property. The most famous classification is that of 1855 which classified the red wines of the Médoc district (together with Château Haut Brion situated in the Graves district) and the sweet white wines of Sauternes.

Cuvée. The word has many different meanings. It can mean the juice from the first pressing for Champagne or a blend and its components of any wine.

Dosage. Following disgorging, the topping up of the bottles of sparkling wine produced by the traditional method (q.v.) with sweetened wine, according to the intended style.

Dry extract. The total of the non-volatile solids in a wine.

Ethanol. Alcohol in wine and other alcoholic drinks, also known as ethyl alcohol (CH_3CH_2OH).

Fining. Removal of microscopic, troublesome matter (colloids) from wine before bottling. Materials that may be used for fining include egg whites, bentonite (q.v.), isinglass and gelatine.

Flash détente. A technique whereby must is heated to 85°C and sent to a high-pressure vacuum chamber where the liquid is vapourised. The skins are deconstructed and the chamber cooled rapidly to 32°C. The process gives well-coloured, fruity red wines with soft tannins.

Formazine turbidity unit. The ISO adopted measurement of fine particles in a liquid by method of light scattering.

Grand Cru. Lit. great growth. A French term for a wine-producing property or vineyard officially rated as superior.

Grand vin. General term for the 'top' wine from a producer, implying that it is the result of selecting the best component wines.

Hectare.   A measure of land, equalling 10,000 m^2 or 2.47 acres.

Kabinett.   A category of German wine, made from ripe but not overripe grapes, without chaptalisation (q.v.), usually light and delicate.

Lees.   Sediment of dead yeast cells. Following completion of fermentation, heavy *gross lees* will fall from the wine, followed by *fine lees* which may not descend until after the first post-fermentation racking.

Malolactic fermentation.   Process of conversion of malic acid into softer lactic acid by the action of bacteria. Although an essential part of the production of red wines, its desirability in whites depends upon the style required.

Master of Wine.   A member of the Institute of Masters of Wine, who has passed the rigorous theoretical and tasting examinations, and has had a dissertation accepted. At the time of writing there are some 278 Masters of Wine worldwide.

Must.   Unfermented grape juice and solids.

Nanogram.   One billionth of a gram.

Oloroso.   A style of sherry that has been matured in casks that are not completely filled, and thus deliberately oxidised.

pH.   A measure of the hydrogen-ion concentration in a polar liquid and, hence the acidity of the liquid. pH is measured on a scale of 0–14, where 7 is neutral, below 7 is acidic and above 7 is alkaline.

Phenol.   Basic building block of a group of chemical compounds that include polyphenols (q.v.).

Picogramme.   One trillionth of a gram: 10^{-12}.

Pipe.   Cask of 534 or 550 L capacity, used in the Douro Valley in Portugal for port production.

Polyphenols.   Group of chemical compounds present on grape skins that includes anthocyanins (q.v.) and tannins (q.v.).

Racking.   Transferring juice or wine from one vat or barrel to another, leaving behind any lees or sediment. The process helps to clarify wine.

Saignée.   Lit. bled. A method of producing rosé wines (and concentrating the colour of reds). Juice which has acquired the desired amount of colour from maceration upon red grape skins is drawn from a vat after a period of 4 to 24 hours, and continues fermentation like a white wine, that is off the skins.

Tannins. A loose term encompassing polyphenols that bind and precipitate protein. Present in grape skins, stalks and seeds (condensed tannins) and also oak products (hydrolysable tannins). Tannins have an astringent feel in the mouth, particularly on the gums.

Teinturier. Vines producing grapes with red flesh, sometimes used to add colour to pale wines; Alicante Bouschet and Dunkelfelder are two of the best known.

Traditional method. The method of producing high-quality sparkling wines and Champagne, utilising a second fermentation in bottle and the eventual removal of yeasty sediment by a process of riddling and disgorging.

VDN (vin doux naturel). Lit. naturally sweet wine. A type of French wine that has had the fermentation arrested by the addition of grape spirit leaving considerable residual sugar.

Veraison. The beginning of grape ripening at which the skin softens and the colour starts to change. Following veraison the growth in the size of the berry is due to the expansion rather than the reproduction of the cells.

Vin Jaune. Lit. yellow wine. Wine that has been aged in cask for several years without being topped up, resulting in oxidation. A speciality of the Jura region in France.

Volatile acidity. Present in all wine, resulting from the oxidation of alcohol to acetic acid (the acid in vinegar). Small amounts enhance aroma but excessive amounts cause vinegary smell and taste.

Bibliography

Bartoshuk, L. (1993) Genetic and pathological taste variation: what can we learn from animal models and human disease? In: *The Molecular Basis of Smell and Taste* (eds, Chadwick, D., March, J. and Goode, J.), pp. 251–267. John Wiley & Sons, Chichester.

Bird, D. (2005) *Understanding Wine Technology*, 2nd edn. DBQA Publishing, Nottingham.

Castrioto-Scanderberg, A., Hagberg, G.E., Cerasa A., *et al.* (2005) The appreciation of wine by sommeliers: a functional magnetic resonance study of sensory integration. *Neuroimage*, 25, 570–578.

Chatonnet, P., Dubourdieu, D. and Boidron, J.N. (1995) The influence of *Brettanomyces/Dekkera* sp. yeasts and lactic acid bacteria on the ethylphenol content of red wines. *American Journal of Enology and Viticulture*, 46, 463–468.

Clarke, R.J. and Bakker, J. (2004) *Wine Flavour Chemistry*. Blackwell Publishing, Oxford.

Crossen, T. (1997) *Venture into Viticulture*. Country Wide Press, Woodend, Australia.

Fielden, C. (2005) *Exploring the World of Wines and Spirits*. Wine and Spirit Education Trust, London.

Goode, J. (2005) *Wine Science*. Mitchell Beazley, London.

Grainger, K. and Tattersall, H. (2005) *Wine Production Vine to Bottle*. Blackwell Publishing, Oxford.

Halliday, J. and Johnson, H. (2006) *The Art and Science of Wine*, 2nd edn. Mitchell Beazley, London.

Hanson, A. (1982) *Burgundy*. Faber, London.

Iland, P., Bruer, N., Edwards, G., Weeks, S. and Wilkes, E. (2004) *Chemical Analysis of Grapes and Wine: Techniques and Concepts*. Winetitles, Adelaide.

Iland, P., Grbin, P., Grinbergs, M., Schmidtke, L. and Soden, A. (2007) *Microbiological Analysis of Grapes and Wine: Techniques and Concepts*. Winetitles, Adelaide.

WINE QUALITY

Michelsen, C.S. (2005) *Tasting & Grading Wine*. JAC International, Limhamn.

Nicholas, P. (ed.) (2004) *Soil, Irrigation and Nutrition*. South Australian Research and Development Institute, Adelaide.

Parker, R. (2005) *The World's Greatest Wine Estates*. Dorling Kindersley, London.

Penning Rowsell, E. (1979) *The Wines of Bordeaux*, 4th edn. Penguin, Harmondsworth.

Peynaud, E. (1987) *The Taste of Wine*. John Wiley, New York.

Rankine, B. (2004) *Making Good Wine*. Macmillan, Sydney.

Robinson, J. (ed.) (2006) *The Oxford Companion to Wine*, 3rd edn. Oxford University Press, Oxford.

Saintsbury, G. (1920) *Notes from a Cellar Book*. Macmillan, London.

Schuster, M. (1989) *Understanding Wine*. Mitchell Beazley, London.

Seguin, G. (1986) "Terroirs" and pedology of wine growing. *Experentia*, 42, 861–871.

Smart, R. and Robinson, M. (1991) *Sunlight into Wine*. Winetitles, Adelaide.

Waldin, M. (2004) *Biodynamic Wines*. Mitchell Beazley, London.

Wilson, J. (1998) *Terroir*. Mitchell Beazley, London.

Useful Websites

This list contains a selection of websites that the author considers to be well constructed and containing valuable information for the reader. It is far from exhaustive.

Association of Wine Educators: http://www.wineeducators.com
The Association of Wine Educators is a professional association whose members are involved in the field of wine education. The website includes a directory of members, many of whom are specialists in different aspects of wine production and assessment.

The Australian Wine Research Institute: http://www.awri.com.au
The Australian Wine Research Institute, whose aim is to advance the competitive edge of the Australian Wine Industry, conducts research into the composition and sensory characteristic of wines, undertakes an analytical service and provides industry development and support. The website contains a good deal of free information.

Blackwell Publishing: http://www.blackwellpublishing.com
The website of this book's publishers, which includes details of publications and links to numerous resources.

CIVB – Conseil Interprofessionnel du Vin de Bordeaux (professional section): http://www.bordeauxprof.com
The CIVB represents, advises and controls the wine industry of Bordeaux, which is the largest fine wine region in the world. This section of their website (in French) includes 'Les Cahiers Techniques du CIVB' – notebooks containing practical information to assist growers and winemakers.

Decanter: http://www.decanter.com
Decanter is a monthly wine magazine, aimed at consumers but widely read by the wine trade and wine producers.

The Drinks Business: http://www.thedrinksbusiness.com

The Drinks Business is a monthly magazine that focuses upon business aspects of the alcoholic drinks industry. There are comprehensive reports on wine regions, trends, brands and in-depth discussions of challenges.

Grainger, Keith: http://www.keithgrainger.com

This website contains contact information about the author of this book.

Harpers: http://www.talkingdrinks.com

Harpers is a weekly journal for the UK wine trade. The website includes précis of recent articles.

The Institute of Masters of Wine: http://www.masters-of-wine.org

The qualification of Master of Wine is regarded as the highest level of achievement in broadly based wine education. The website includes a list of the Institute's members.

ITV – Centre Technique Interprofessionnel de la Vigne et du Vin: http://www.itvfrance.com

ITV provides technical information for growers and winemakers in France. The website (in French) includes details of their publications.

Lincoln University Centre for Viticulture and Oenology: http://www.lincoln.ac.nz

Lincoln University, based at Christchurch, New Zealand, is a centre of excellence for studies and research in viticulture and oenology. Searching under 'wine' on the website links to courses and staff members.

New Zealand Winegrowers: http://www.nzwine.com

The New Zealand Winegrowers website contains valuable information about the industry, including statistics, profiles, reports and links to producers.

UC Davis: http://wineserver.ucdavis.edu

The Department of Viticulture and Enology at Davis has been at the forefront of research for over 30 years.

Uiniversité Victor Segalen Bordeaux 2, Faculté d'Oenologie: http://www.oenologie.u-bordeaux2.fr/

Centre of excellence for studies of oenology in France.

The Wine Anorak: http://www.wineanorak.com

A lively online wine magazine produced by Jamie Goode, who unusually for wine writers has a strong scientific background.

Wine Australia: http://www.wineaustralia.com

A valuable source of up-to-date news and statistics of the Australian wine industry.

Wines of Chile: http://www.winesofchile.org

Wines of Chile represents 90 Chilean wineries. The website is a useful source of news, statistics and reports.

Wines of South Africa: http://www.wosa.co.za

Wines of South Africa represents exporters of South African wines. The website contains up-to-date news and links to useful statistics.

Wine & Spirit Education Trust: http://www.wsetglobal.com

The Wine & Spirit Education Trust designs and provides wine courses, and is the industry awards body whose qualifications are recognised by the QCA in the UK under the National Qualifications Framework.

The organisation is international and there are approved programme providers in over 35 countries worldwide.

Winetitles: http://www.winetitles.com.au

Winetitles is an Australian publisher of wine books, journals and seminar papers.

Wine Exhibitions

Listed below are some of the most important wine exhibitions.

United Kingdom

LONDON INTERNATIONAL WINE FAIR
Held in London at the Excel Exhibition Centre in Docklands on Tuesday, Wednesday and Thursday, normally in the second or third week of May. The fair is a trade only exhibition.
www.londonwinefair.com

France

VINEXPO
Held in Bordeaux biannually in June. This is huge, trade only exhibition.
www.vinexpo.fr

Germany

PROWEIN
Held in Düsseldorf over 3 days in mid-March. This trade only exhibition has gained in importance in recent years.
www.proweind.de

Italy

VINITALY
Held in Verona over 5 days in early April, this trade only exhibition is by far the most important Italian wine event.
www.vinitaly.com

Spain

FENAVIN
Held in Ciudad Real in May, this recently established exhibition focuses on the wines of Spain.
www.fenavin.com

China

VINEXPO
Asia Pacific is held in Hong Kong over 4 days in late May.
www.vinexpo.fr

Index